Once a Fighter Pilot . . .

Once a Fighter Pilot . . .

Brig. Gen. Jerry W. Cook
USAFR, Retired

McGraw-Hill
New York Chicago San Francisco Lisbon London Madrid
Mexico City Milan New Delhi San Juan Seoul
Singapore Sydney Toronto

Cataloging-in-Publication Data is on file with the Library of Congress

McGraw-Hill 

A Division of The ***McGraw-Hill*** *Companies*

2 3 4 5 6 7 8 9 0 DOC/DOC 0 9 8 7 6 5 4 3 2

ISBN 0-07-139920-8

The sponsoring editor for this book was Shelley Ingram Carr and the production supervisor was Sherri Souffrance. It was set in Garamond per the Tab3 design by McGraw-Hill's Blue Ridge Summit, PA, Professional Book Group composition unit.

Printed and bound by R. R. Donnelley & Sons Company.

This book is printed on recycled, acid-free paper containing a minimum of 50% recycled de-inked fiber.

Contents

To the guys in the "jungle."

I was a "fast mover."
I rarely got a glimpse of you as I roared by,
but I never forgot that you were there.

You are the real heroes,
especially those of you on the Wall.

Acknowledgments

I wish to acknowledge friends and family members whose gentle nudging and warm encouragement helped to bring this work to reality. Luraette Tucker, the first nonfamily member to request and read the original manuscript. Joyce and Jim Faulkner, who early on expressed enthusiasm for the book. Andy Holbert, a pilot buddy who read some of the first stories and said, “Keep writing.” Russ Johnson, whose experience and advice helped me to structure the text. Becky Kossover, whose red pen and expertise helped keep me on the right track.

The Coffields, Ellen and Jim, who “critiqued” with kindness. Leah Williams, a friend whom I've never met, for her enlightening questions, comments, and kind words. Jack Stephens, for the use of his personal time and valued opinion. Waymon Pearson, an old friend and fellow pilot, for his expert viewpoint.

The Scroggins, Vearn and Johnny, for their enthusiastic responses to the manuscript. Mary and Greg Hubbard for their gratifying requests for “more, more” stories. The Johnsons, Elsie and Tommy, for their resolute support.

One of the bravest men in Vietnam, my regular “backseater,” Vern Hammett, for his coolness under fire during some of my wilder “maneuvers” and his kind response to my recent requests for assistance.

A special thanks goes to Susan Taylor, whose enthusiasm for this project, faith in my ability, and hours of unselfish proofreading, had a very large part in its completion.

Most of all, to the person who started it all. Her idea from the beginning: She never wavered in her encouragement and support. Her unfailing confidence in me is what caused me to try. My deepest appreciation and gratitude go to Linda Ann Cook, my wife and best friend.

Introduction

This book began as a project suggested by my wife, Linda. She thinks that there comes a time in almost every life when one feels a desire to know more regarding relatives, especially ancestors about whom there is very little information.

I have an old tintype that my mother gave me of an ancestor who served in the Civil War. He appears in uniform, and the family resemblance is remarkable; however, the only other reference I have seen about him is a medical discharge. I am fascinated, but I will never know more. Everyone who could have told me of him has passed away. The knowledge is gone forever. Perhaps he was not an important person to anyone else, but I still have this innate desire to know about him. I never will.

Linda thinks that someday a grandchild or some other descendant might feel the same way about me. She thinks that he or she might be interested in certain aspects of my life. She has heard me relate some of my experiences to others and thinks that my grandchildren should hear them. It appears that they will never hear them in the usual way, by sitting on their grandfather's lap; hence, *Once a Fighter Pilot.*

These stories are based on memories of events that occurred thirty or more years ago. Although I have strived to recall all particulars as accurately as possible, certain details have been "lost" forever. Somewhat like a woodland place once visited, nuances are now grown over by time, never to be found again; however, even though some of the specific aspects have gone, the emotions and "feel" of the place will always remain. They are herein faithfully recorded.

Kid's dream

Bartlesville, Oklahoma, 1942

The Lockheed P-38 was absolutely beautiful! The sound from the twin V-12 engines was incredible. It was 1942, and I was five years old. I didn't know how he had found me, but I was convinced that the fighter pilot was doing the rolls and loops over my backyard just for me and me alone. You could never have convinced me otherwise.

He has no idea the impression he made that day, or the lasting effect that his flying created. I don't know who he was or whether he survived the war. I don't know who the "airshow" was really for, probably a girlfriend or his parents. Perhaps he is still alive and reading this. If so, my heartfelt thanks goes to him. From that day, I have lived and breathed flying, particularly as a fighter pilot. If he is still alive, he knows what I mean. "*Once a fighter pilot,* always a fighter pilot."

Where do I sign?

University of Oklahoma, 1956

My mother wanted me to be a doctor. I had always known she wanted me to be a doctor. When I mentioned airplanes, we would end up talking about medicine. There had never been a doctor in our family. Medicine was the pinnacle of professions. There had never been a pilot in our family either, but that somehow was not a valid point with her. Somehow it just was not the same. I probably should have listened and heeded her wishes, but I wanted to be a pilot. Sitting on the back row in chemistry class, I wanted to be a pilot. Staring out the window in French class, I wanted to be a pilot. In zoology class, dissecting frogs or slicing open a sheep's eyeball, I *especially* wanted to be a pilot!

I was an only child, and I felt a particular responsibility to my parents. There were no siblings with which to share my parents' dreams. I never wanted to fail them. I was all they had. That must have been a burden for them! I don't know why they never had any more, and I never asked. Maybe they didn't want to take any more chances after what they ended up with the first time. In any case, I knew and appreciated the sacrifices they had made, and were making, to pay for my college education. I had been a pretty good student in high school. Without having to study much, I had made good enough grades to belong to the National Honor Society. I had also been the president of both the junior and senior classes, so I thought I was pretty "cool" and smart. Was I in for a rude awakening! I was a small duck in a huge pond of fifteen thousand or so other "cool" smart students. I remember thinking that they all seemed a lot "cooler" than I. As far as smart, I was struggling in French, flunking chemistry, and loathing zoology. I felt so guilty. I was totally wasting my parents' hard-earned money. The only class in which I was doing well was Air Force ROTC. That one I was tearing up. In retrospect, it's interesting to me how my mind operated. Consciously, I was trying in the other classes, but to no avail. Subconsciously, almost without trying, I was succeeding in a subject because it held my interest.

The recruiting sergeant was wearing his dress uniform with all his ribbons. As he talked to us about the Air Force in general, I listened, but without much enthusiasm. He was discussing various career fields available to ROTC graduates. All of them were ground jobs, supply, electronics, personnel, transportation, and the like. They all sounded boring and tedious.

"Last, let's talk a little bit about the reason the Air Force even exists," he said. I sat up straight and focused all my attention on his words. "Flying is what all other Air Force specialties are there to support. Directly or indirectly, every job in the Air Force only exists to help the planes and pilots fly and fight," he said. Now he had my attention. He talked about the mental and physical requirements to be a pilot. He mentioned that as a college and ROTC graduate that we could apply for and take the "stanine" tests. If he passed the stanine, an applicant would then receive the standard Air Force flight physical. If he passed that, he would be considered for USAF pilot training.

He then began the statement that would change my life. "There is also a program for noncollege graduates. It's called the Aviation Cadet Program. If anyone here is interested in hearing more, see me after class."

"Where do I sign and when and where is the test?" were the first words tumbling out of my mouth. Five minutes later, I was taking the first steps toward fulfilling my childhood dream.

Stanine was the type of examination that I was now taking. I had never heard of this kind of test before, but it was a bear. It took about eight hours to complete, and when I finished, I didn't have the vaguest idea whether I had passed or failed! After spending that night in the barracks at Tinker AFB, I reported the next morning to the testing classroom. The list of those who passed was to be posted outside the door with directions to the base hospital for flight physicals. The list of the failures was also posted, along with a "Good luck in another career" message. The pass list was short. Of the twenty or so who had tested, only three had passed! I didn't have a clue how, but there I was in one of those little hospital gowns, getting half of my vital fluids drained and trying, in vain, to keep the back half of my gown closed.

For some reason, the only thing worrying me was the eye exam. I had heard that you had to have twenty-twenty vision to be an Air Force pilot. I had never had an eye exam before. I had just been given that particular test a few minutes earlier and was sweating bullets. I was standing by the nurses' station near the spot where our physical exam paperwork was stacked awaiting the next phase. My paperwork was on top! I didn't know if I was supposed to look, so I was very careful as I stole a long glance. I saw it. I was devastated! My eyes had tested twenty-ten, instead of twenty-twenty. I went through the rest of my physical in a crestfallen mood. I knew that in the last phase of the physical, we were examined by a flight surgeon. He would review the tests, see the eye results, and I'd be getting my walking papers.

"Mister Cook, your physical results look excellent. We can't find a thing wrong with you," the doctor was saying. He had missed it! He had missed it, and I sure wasn't going to tell him. I couldn't believe my good fortune; however, I immediately began almost two years of concern that someone would catch the error and I'd be finished. Ignorance can certainly cause a lot of needless worry, I later learned.

My parents didn't have any idea when I walked through the door. They thought that I was just home for semester break. They didn't know that I had dropped out of college and was home until the United States Air Force called me. They weren't happy campers, but I admit that they took the news better than I probably would have.

My parents owned and operated a Pontiac, Buick, and GMC dealership. I went to work for them and began my wait for the good news. I had taken the test in January and it was now nearly spring. One day a friend of mine from high school came into the dealership to visit. He was home on leave from the Air Force. Although he was several years ahead of me, our school was small and everyone knew everyone else. He had become an Air Force pilot, and I was eager to talk to him about it. He had gone the Aviation Cadet route also and could tell me some of what to expect. One of the first things he said nearly floored me. He related that not everyone who passed the tests was accepted. Passing the tests only allowed you to be considered. You were selected based on how well you had scored in comparison with everyone else and on a review of your physical. I figured that I had really had it now. To add to my consternation, he proceeded to tell me that only three percent who took the stanine nationally passed. Then only about ten percent of those passed the physical. Those were the guys I was competing with and I had *bad* eyes. When he left, my confidence level was zero.

I didn't tell my parents of my fears. Eventually I knew I had to tell them, but not yet.

When the envelope came for me in May, I still hadn't told them. When I saw the return address, I figured the day had arrived. It was from the United States Air Force. As I slowly opened the envelope, my hands were shaking. I had to read it, but I didn't want to. As long as I didn't know for sure, there was still hope, right? I couldn't believe it! Instead of washing cars for my folks or being back in college, I would be at Lackland AFB, Texas, on July 3, 1957. I was accepted for Pilot Training Class 59-C in the United States Air Force. Nearly forty years later I came across the letter I received that day. I could almost recall the excitement I had felt. I had to smile to myself remembering how naive and youthful I was.

It is definitely a blessing that we can't see the future. I doubt we would even attempt half of what we accomplish.

Welcome to the Air Force. Now, forward march!

Oklahoma City, Oklahoma, July 1957

I had never been in an airplane before. Here I was, heading to pilot training and I had never set foot inside an airplane! "Central Airlines flight to Dallas and San Antonio now boarding" came the announcement. I hugged my mother and shook my dad's hand. They looked very sad and concerned. Me, I was excited and scared at the same time. I wasn't afraid of flying, but of failing. Whatever I did, I couldn't fail. I had dreamed about this too long. I wouldn't fail! Now began a time in my life that I wouldn't trade, but I'm not sure that I'd go through it again.

The flight to San Antonio was too short. I was fascinated at the sensation of flying. Looking at the ground, houses, and cars reminded me of toys. It was to be the longest twelve weeks of my life before I would see inside another airplane. We landed at San Antonio International Airport, and I walked inside the terminal. I spotted an information desk manned by an Air Force sergeant. He directed me to a door and beyond that a blue Air Force bus with Lackland AFB on the front. I stepped on the bus not realizing how drastically my life was about to change.

"What's your name mister?" yelled the sergeant, right in my face.

"Jerry Cook," I replied.

"Wrong!" he yelled again. "You are New Aviation Cadet Cook, mister," he said, "and if you don't wipe that smile off your face, you'll be called New Aviation Cadet Blithering Idiot," he yelled again.

I remember thinking, "I'll be deaf New Aviation Cadet Cook if you keep yelling." I struggled to keep from smiling as he continued to chew on me and the other cadets standing beside me at our amateur rendition of attention. This was just like the Air Force pilot had told me it would be. I found out that he hadn't told me quite everything. I soon had no trouble at all keeping from smiling. I was too tired.

Cadets ran everywhere for the first two weeks. After that, we sometimes marched between our runs. Reveille would blow at 5 A.M. In five minutes we had to be outside for roll call and calisthenics. By five-thirty we were in the shower and by five-forty five fully dressed and marching to breakfast. We had to eat looking straight ahead. We had to sit on the forward three inches of our chair at attention. We couldn't drop anything. We couldn't spill anything. The only thing I saw was the bald-headed guy opposite me. The only thing he could see was the bald-headed guy opposite him, me! It wasn't a pretty sight.

The Technical Instructors, or TIs, and the upper class were merciless if you dropped anything. I decided it wasn't worth the hassle to eat. From that first day I learned not to get anything on my tray but milk because you had to eat everything that you selected. I'm surprised that I still like milk. I survived on it and hardly anything else for the first four weeks. The pressure then let up somewhat because the new class had arrived; however, we still marched or ran everywhere. We were now in superb condition, mere shadows of our former selves. Our hair had even grown back a little, making our ears seem somewhat smaller.

Academics: When I wasn't marching, running, or doing calisthenics, I was studying. I thought college was hard. If I had studied like that in college, I could have been a doctor or anything else that I wanted. But I had wanted this, and I still wanted it. Even with all the pain, all the harassment, all the grief, I never lost sight of my dream. I resolved not to let anything keep me from my goal of becoming a USAF fighter pilot. My biggest obstacle was my temper.

It was Saturday morning. It was inspection time. We were billeted two to a room. Our room was spotless. It had to be. The inspecting officer wore white gloves. There couldn't be a speck of dirt anywhere. If there was, we would be walking on the "tour ramp," one hour for each demerit. Free time was precious little, and I didn't want to spend it walking in circles for hours. I had been at Lackland for eight weeks and was proud that I hadn't received enough demerits to walk any tours yet. That was somewhat unique. Very few cadets got through the program with no tour walks. I wanted to be one of the few.

Our upper class was billeted upstairs from us. They were to graduate in about one week, then we would become the upper class. With that came a few more privileges. Also, it would be much easier to avoid demerits with them gone. One of the upper-class cadets had evidently decided that I would not escape the tour ramp any longer. Just as we heard the inspecting officer enter the barracks at the far end, that particular upper-class cadet appeared. We were standing at attention at the foot of our bunks. We were supposed to remain at attention when an upperclassman was in the room. We did. Although I was looking straight ahead, in my peripheral vision I could see him moving about by my bed. In the meantime, I heard room after room being called to attention as the inspector went in and out of them at the far end of the barracks. They were getting closer. Suddenly the upperclassman "tigered" my bunk. That means he ripped the blanket and sheet from it and tossed it on the floor.

He stepped back and said, "Cadet Cook, your bunk wasn't tight enough. Fix it." The blanket on our bed had to be tucked in so tight that a quarter dropped on it would bounce two or three inches high. I knew that mine had been made properly. I had given it the quarter test just a few minutes prior. I quickly made the bed again as the inspecting officer got closer. I checked it with the quarter and hurriedly returned to my position at attention. Thinking he would now leave, I couldn't believe it when the cadet again tigered my bunk. I was rapidly running out of time. I made the bunk again as the upperclassman watched with a smile on his face. The inspection team was much closer.

I straightened up with a red face and said, "You will not tiger my bunk again, Sir!" He had leaned over and grabbed the blanket from my bunk for the third time when I hit him. He landed in the hallway with a thud and grabbed his nose. It was bleeding profusely. I returned to my position of attention. He quickly stood up and pointed at me.

"You're finished Mr. Cook! I've finally got you. You can't hit an upperclassman and get away with it," he sputtered.

"He can this one time, Mister." The Cadet Wing Commander stepped around the doorjamb of my room. "I heard the whole thing," he addressed his classmate. "You are way out of line for even being down here just before an inspection, much less doing what you did," he continued. "Now get back to your own room before I decide to take further action against you." The wounded upperclassman left still holding his nose.

"Mr. Cook, I can't officially condone what you just did. You're very lucky that I was right outside and heard everything. Off the record, I'd say that he deserved exactly what he got. Don't worry, he will take no action against you or bother you again. But remember what you've been taught, 'discretion is the better part of valor.'" I have remembered. I can't say that I've always followed the precept, but I have remembered.

We passed the inspection. That particular upperclassman never came around me any more. I finished the Aviation Cadet program at Lackland without walking a tour. If the harassing upper-class cadet had gotten his way, I wouldn't have finished at all. I was lucky. I shouldn't have hit him. I almost lost my dream because I lost my temper. In a fighter plane, it could have cost my life.

In later years, one of my friends forfeited his life because he lost his cool while flying a high-performance fighter. What a terrible price to pay!

Finally, airplanes

Hondo Air Base, Texas, October 1957

It seemed like there were airplanes everywhere. As we drove through the gate, I felt like I had died and gone to heaven. Actually, I had only driven about thirty miles west of Lackland. I was in my 1949 blue Ford coupe with a couple of other cadets who didn't have cars. We had graduated from Preflight Training and were finally reporting for Primary Flight Training. The T-34 Mentors and T-28 Trojans buzzing around the sky made my heart pound as if a beautiful girl was smiling at me. I sure liked pretty girls but right now all I could think about were these beautiful airplanes in every direction we looked.

"Man," I thought. "Now I'll fly!" Wrong! First we marched! That is, when we weren't running somewhere. Welcome to the real world of Aviation Cadets. We were again the underclass. *Everyone* outranked us. We found out that our rank structure was below the lowest ranking airman. I knew our pay must be the lowest, but now we were informed that so were we. Oh well, someday we would be second lieutenants, and we thought they were right up there next to "you know who." But first, academics!

I found out that we wouldn't even get to touch an airplane before weeks of academics. The only thing more forbidden for us to touch was girls! But they weren't flying around all over the place to tempt us like the airplanes were. If there had been any girls around, and there weren't, who had the time or the energy? The 5 A.M. reveilles and calisthenics had followed us the thirty miles from Lackland.

I'll never forget my first flight at the controls of a T-34. We were to fly the Mentor about thirty hours, then move into the T-28 which was a much larger and slightly faster plane. The flying program at Hondo Air Base, Texas, was operated by a company called Texas Aviation Industries. It was a civilian concern under contract to the Air Force. All the instructors, except for a few military check pilots, were civilians. My instructor's name was Mr. Kurkendal. I can remember exactly what he looked like after thirty-seven years. I can also remember his personality. He was kind, calm natured, and very patient. He never raised his voice or became angry. Some of my classmates were not so fortunate. Several of them had problems during the training that, I think, were directly related to the berating and unnecessary pressures from their instructors. I determined that if I ever became an instructor, I would try to emulate Mr. Kurkendal. I eventually graduated at the top of the flying phase in my class, and I attribute it all to my instructor. He was the best.

My instructor climbed in the back seat of the little T-34 after ensuring that I was properly strapped in up front. He checked in on our interphone headsets and talked me through the engine start. We had thoroughly briefed the flight beforehand, so it went relatively smoothly. We taxied slowly to the runway, making some practice turns and stops along the way. The mobile control tower of the T-34 runway cleared us for takeoff. I can't remember the takeoff at all, except for feeling Mr. Kurkendal lightly on the control stick with me. He talked me out of the traffic pattern, and we went for my "dollar ride" as it was called. We looked over the T-34 flying area and made a few turns, climbs, and descents. We lowered the landing gear and flaps and simulated a landing pattern. Then we headed back to the base for the real thing. It hadn't lasted nearly long enough, but my instructor had three other students waiting with their dollars ready. Calmly he talked me through the approach and landing. I do remember my first landing very well. It really wasn't that bad. I couldn't feel my instructor on the controls, though it's certainly possible that he was. As we stood at the trailing edge of the right wing, Mr. Kurkendal showed me how to fill in the flight records form. We had only flown for eighteen minutes. It had seemed to me only like five or ten. Then he asked me, what I thought at the time was a strange question. "How much flying time do you have, Mr. Cook?" he asked.

I looked down at the form we had just filled out and replied, "Eighteen minutes, sir."

"No, I mean how much total flying time do you have?" he said.

I thought it must be a trick question or something, but I answered it again with "Eighteen minutes, sir."

I couldn't read the funny look on his face, but I knew things must be okay when he grinned and said, "Eighteen minutes, huh, how interesting." I hadn't realized at the time that there were several students who already had a civilian pilot's license. In fact, I later found out that one of them had flown over three thousand hours as a civilian. He was good!

I was now a USAF student pilot. I knew that academic and officer training were important, but I couldn't wait to get back in the air each time. It seemed like I blinked and T-34 training was over. It was a fun little airplane, but I was ready to tackle something bigger and better, the T-28.

We had completed the initial transition phase and were now entering a completely new stage, night flying. After a couple of dual night flights with my instructor, he felt that I was ready to try it solo. What happened next I feel warrants its own story. It was almost my final chapter.

God's country

Hondo Air Base, Texas, 1957

"This is God's country. Don't drive through it like Hell," read the sign outside Hondo, Texas. I don't know whether the sign is still standing, but it and John Wayne almost bracketed my final resting place.

I probably had a grand total of about sixty hours flying time, thirty of which was in the little T-34 Mentor. The T-28 seemed three times its size. The big radial engine swung a very long two-bladed propeller. The engine was cranky and hard to start when it was cold but seemed very reliable otherwise. There was a primer button to enrich the fuel mixture, which assisted in starting. That little button became a big part of my life that night.

All was normal during the engine runup. The engine gauges were in the right regions, and it sounded as smooth as the big radial ever got. I was cleared for takeoff. It was really dark. I was excited and a little nervous as I added rudder to keep the nose tracking straight down the runway. I eased the stick back at the proper speed, and the T-28 lifted into the night air. I raised the landing gear handle and gained some more altitude prior to retracting the flaps. Just as I moved the flap handle to the up position, I heard the loudest noise of my young life, sudden and near complete silence! All I could hear was the wind noise around the canopy and my own heart beating in my ears. As I began losing altitude, I turned slightly right where I knew there were some open fields and to avoid the local drive-in theater. As I fumbled in the darkened cockpit for the engine control panel, I saw the drive-in screen lit up below and to my left. Funny the things that flash before your eyes at a time like that. How about John Wayne and the *Flying Leathernecks* for instance!

I pressed the primer button and the engine caught and began running, albeit roughly, but running. As I let go of the primer button to try to answer the control tower's slightly frantic sounding calls to me, the engine started sputtering again. I quickly held the primer button, and it started again. The tower continued calling me, but there was no way I was going to let go of that button again. I felt that it was time to establish priorities, you see.

I gingerly turned the rough-running T-28 back toward the runway. The engine was producing partial power, and I was having trouble maintaining altitude. I decided to take a chance on a landing in the opposite direction to which I had just taken off. I didn't see aircraft lights on the runway, and I could hear the tower breaking everyone else out of the traffic pattern. I also heard them canceling someone's clearance to taxi onto the runway.

I wasn't about to quit holding the primer long enough to put the gear handle down, so I let go of the control stick and slapped it down with that hand. The control tower saw my landing lights and cleared me to land on any runway. After the landing, of which I recall little, I turned off the runway and finally let go of the primer button. You guessed it, the engine continued to run! The fire trucks surrounding the plane kept me from taxiing, so I shut down the engine.

The next day, I was told that the mechanics couldn't duplicate the engine problem. They surmised that perhaps the constant priming had pushed some trash out of the induction system.

I think perhaps my landing, which I don't remember, may have knocked the trash out! One of my friends, who witnessed it, said it wasn't very pretty.

I think John Wayne was flying again the next night when I took off, but I can assure you of one thing. His engine was running a lot smoother than I imagined mine to be!

The boss's daughter

Hondo Air Base, Texas, 1957

To the northeast of Hondo Air Base was a beautiful little body of water called Lake Medina. It was formed on the river of the same name running down out of the Texas "Hill Country," as it is called. It was just out of the north edge of our T-28 transition training area. We were never supposed to stray out of our designated flying zone. To do so was one of the "cardinal sins."

One day a friend in my aviation cadet class was flying a solo transition sortie near the north edge of the area. As he droned along, he was admiring the beauty of the lake and the hill country surrounding it. Then the only thing in the world that could probably make a cadet break a rule raised its pretty head, "Sex."

As he checked out some of the houses along the lakeside, he saw blonde hair, a small bathing suit, and long, long legs. Even from that altitude and distance she looked beautiful; therefore, his course of action was obvious. The poor guy didn't have a choice. He *had* to get closer. There wasn't any question about it. He didn't have any binoculars! He had to roll in and go investigate. This might be someone he would want to come looking for in his car on his next weekend pass. Besides, this would be his chance to show off a little of the flying skills that he had been working so hard to master.

He flipped the T-28 upside down and pulled the nose through. He pointed it just below the "target" and rolled upright again as he leveled off just above the water. He knew that he shouldn't be doing this. He wasn't even supposed to be out of his flying area, and here he was down "on the deck" getting ready to "lay" a buzz job on a fair damsel *not* in distress.

As he roared over the dock, he racked the big propeller-driven trainer up on its side and looked down. She looked even better up close than he thought she would. She was a doll.

She waved and smiled as he swept by. "Wow," he thought, "She likes it. Now I'll really show her something." He sped off across the lake and made a climbing turn. He again pointed the nose at the dock with its pretty occupant. She had stood up now and was shielding her eyes from the sun as he roared in her direction. He descended lower and lower over the water. Just before he got there, he pulled the nose up sharply and did an aileron roll. He had never done one so low before and he was elated. He was getting braver by the minute. He decided to test his skills even further and impress his future date as well.

Once again he pulled the nose around toward the object of his undivided attention. This time, long before he got to the dock, he pulled up the nose and did a half aileron roll. He held the T-28 Trojan upside down. He had to push forward hard on the stick to keep his airplane from descending. As he looked "up" at the approaching dock, he saw that his lovely audience was no longer alone. There was an older gentlemen peering at him thorough a pair of binoculars.

"Uh-oh," he thought. "I'd better get out of here. That's probably her dad, and he's probably trying to get my aircraft number."

The thought of getting in trouble and getting washed out of the cadet program scared him worse than any thoughts of busting his butt doing unauthorized and unsafe flying at such a low altitude. *And* he was out of his transition area.

He rolled the Trojan upright and headed for home base as fast as he could. He was hoping and praying that the gentleman on the dock had not been able to read his number. Maybe he was so close to being directly overhead the dock that they couldn't be seen.

I was waiting in a line of T-28s for takeoff. I saw a single T-28 coming down initial to make his overhead break for landing. As he rolled into his sixty degree left turn to the downwind pattern, his aircraft looked strange. I didn't get a good look at first, so I didn't know exactly what was different about the plane. As he flared and settled down on the runway, I could see clearly the difference. His T-28 was partly black! From near the exhaust stacks rearward, there was a long wide streak of something black all the way back to the tail.

I then realized what it was. Oil! There was black oil everywhere. He didn't have a clue that he was flying a different color airplane than everyone else. He was just worried about the guy back on the dock getting his tail number. He had worried about the numbers needlessly. The numbers were covered with oil, too.

Now as Mr. Paul Harvey aptly states, here's "the rest of the story."

The young lady on the dock was very pretty indeed. She was also the daughter of a man who was fairly important to the aerobatic cadet. He was flying an airplane that "belonged" to the gentleman. Her father happened to be the president of Texas Aviation Industries, which was the company providing the aircraft and instructors for our Primary Flying School.

Several people were waiting for our cadet friend when he parked. He figured that his fears were realized and that the guy had read his number and called the base.

The call had indeed been made. But they didn't need a number. All they had to look for was the only black T-28 on the ramp. It seems the extended inverted flying had dumped a bunch of oil out of its tank and along the side of the Trojan.

Do you suppose that is one reason why inverted flying was prohibited?

No, he didn't get kicked out of cadets. But he didn't have any free weekends to go hunting up the little lady on the dock either.

They probably figured that he was such a good pilot to not have killed himself that it would have been a waste to wash him out of pilot training.

Besides, what healthy young male can resist getting a closer look at a pretty girl!

The Honor Code

Hondo Air Base, Texas, 1958

The "Honor Code." It was the central point around which an aviation cadet's life orbited. It worked very impressively. You could leave anything lying out, and it would be there when you returned. Nothing had to be, or even could be, locked away for safe keeping. It was effective, but it was also merciless. If you violated the Code, you were gone, out of Air Force pilot training immediately. Your flying career in the United States Air Force was over before it even began.

I was in a head-to-head competition with another aviation cadet to graduate number one in our pilot training class at Hondo. He had brought a slight advantage with him when he reported to USAF pilot training. He had accumulated over three thousand flying hours and had been employed as a cropduster pilot. If I had any edge, it was that I didn't have to relearn anything the Air Force way or break any old habits.

Needless to say, I was surprised and pleased at my progress and very determined to excel. After all, I had wanted to be a pilot since I was five years old. My instructor was the best. You may recall that, like all the other Primary Flight Training instructors in those days, he was a civilian. The only military pilots at Hondo were the check pilots. We called them the "hatchet men" because they could cut you from the program. They administered our progress check flights and our final check flights in each flying phase. Occasionally, one would show up and choose a student to fly with that day. We all dreaded that fate because they *were* the hatchet men. We were scared to death of them because they held our future in their hands.

My day came. I was scheduled to fly a primary instrument training flight with my instructor. As I reported for duty at the flight line building, I routinely checked the scheduling board. Sure enough, there was my name in the back seat, but not with my instructor. I had been rescheduled to fly with the chief military check pilot! He had a reputation for eliminating the largest number of student pilots from Hondo's pilot training program. I couldn't believe that I was so unlucky.

My instructor walked up and saw the look on my face. "Don't worry, Mr. Cook," he said. "You're doing great. I think he just needs some flight time before the end of the month. He probably heard how good you are and wants to see for himself," he finished. My mouth was so dry that all I could do was shake my head.

I looked across to my briefing table and there he sat, big, red-headed, and mean-looking. The other students at the table appeared as if they were trying to be invisible. They were all looking down with their heads cradled in their hands. I guess they were afraid he would change his mind and fly with one of them. He was carefully going over my grade book. As I reported to him in the proper military manner, he gave not one hint of kindness or friendliness. That Air Force captain was all business. You must understand that as an aviation cadet I thought that even a second lieutenant sat at the right hand of, well, maybe that guy!

Finally, the briefing was over, and we were in the air. I had accomplished several of the basic instrument maneuvers and felt confident they had gone well. Suddenly, without warning, all the electrical power went off in the T-28 trainer. I was in the back seat beneath the instrument hood, which was a training device to simulate flying inside clouds when you can't see the ground. Attached to the inside of the canopy, it allowed viewing your instruments but prevented seeing outside the airplane unless you lifted its edge. Turning off the electrical power demonstrated what would most likely happen in a real situation if your instruments failed for any reason while in clouds. You would eventually enter a graveyard spiral, which is a steep turning dive at high airspeed.

I couldn't let that happen. Somehow, I had to show this guy how good I was. Something just above my head caught my attention. The sun was shining down through the canopy across a metal canopy support and casting its shadow onto the hood. I moved the stick slightly to one side, then back. The shadow moved, then moved back to its original position on the opaque instrument hood.

"That's it," I reasoned, "Screw the attitude and heading indicators. I'll just hold that shadow in the same place and check my altitude from time to time." It worked like a charm. There I sat with a big grin on my face thinking, "Boy, am I smart."

After several long minutes of flying along like that, the electrical power suddenly came back on and the check pilot yelled, "I've got it. Come out from under that hood, you damned cheater!" I started to say something in explanation, and he yelled "Shut up, don't say another word," and I didn't.

The "Honor Code," no stealing, no lying, no cheating. One violation and you were gone!

We raced back to the air base, and he landed the aircraft. As he unstrapped and turned around, his eyes were blazing. "You're a cheater, Mr. Cook. There's no place for cheaters in this Air Force," he said.

"But sir," I started.

"I said shut up, mister," he yelled! Tears were welling up in my eyes. I was twenty years old, but I couldn't help it. I tried hard to stop them, but to no avail. I finally made it into the flight shack and hung up my parachute.

My instructor was watching the check pilot with a troubled expression. As I slowly walked over to the table, the check pilot said, "Sit down and be quiet." He was writing furiously on a grade sheet, in red pencil. When he finally was finished, he looked at me with disgust and turned to my instructor. "Mr. Cook cheated when I turned off the electrical power. He continued to fly straight and level for several minutes after the gyros had spun down. He had to have been looking outside, under the edge of the hood, to do that. That's cheating, and that's an automatic exit from the pilot training program!"

Everyone was listening. My face was red and I felt like I was choking. My world, all my plans were gone! My instructor cleared his throat and looked sadly at me. "Why did you look out from under the hood, Mr. Cook? You knew better than to do that, didn't you?" he asked.

"I didn't look outside, sir," I said with difficulty.

"You had to, you liar," the check pilot said loudly.

"Please, captain, let him finish," my instructor said.

Slowly, I told them what I had done. When I had finished, there was a long silence.

The captain turned to my instructor and said, "Would that work?"

"I don't see why not, and it sure as hell wasn't looking outside," he answered with a huge grin.

The check pilot turned and looked at me. "Son, I apologize, I don't believe that I would have thought of that in a million years," he said. "I'd say you not only weren't cheating by looking outside, but you're a hell of a quick thinker." He grabbed my "pink slip" and tore it into several pieces. He then filled out a new grade sheet reflecting an excellent overall. He stood up, shook my hand and said, "Well, I learn something new every day."

I remember thinking, "So do I." And, I'll never forget it!

Merry Christmas

Kingfisher, Oklahoma, 1957

My "Smitty" mufflers were popping as I decelerated from "cruise speed" down to near the posted limit. I was on U.S. Highway 81 on the south edge of my present home town of Kingfisher, Oklahoma. I had lived there since moving from Enid, Oklahoma with my parents when I was in junior high. This was where I had graduated from high school in 1955. I was a Kingfisher High "Yellow Jacket" alumnus.

My parents were perfect examples of the "American Dream." Dad had quit a good job working for an automobile dealer in Enid and moved to Kingfisher to open his own business, Cook Auto Parts. It was very hard work, long hours, and lots of worrying. My mom and dad basically laid it all on the line and went for it. I don't remember that they ever had any employees helping them at the auto parts store. The two of them did it all. They rapidly built a reputation for honesty and fairness and the business was doing well. An opportunity then presented itself. The building next to them was a farm tractor and implement dealer; the dealer had decided to retire and asked them if they wanted to buy his building. After long hours of discussions, planning, and more worry, they again "went for it."

They moved all of their inventory into the next-door building and at the same time closed a deal with General Motors to become Kingfisher's Pontiac, Buick, and GMC truck dealer. Again their integrity and the long hard hours were beginning to pay some "dividends." I certainly had not been any help to them in their American dream, because I was off in the United States Air Force chasing mine.

I pulled my little blue 1949 Ford coupe into a parking spot on the east side of their dealership and walked in. I was home on Christmas leave from Hondo Air Base, Texas. It had been a few months since I had seen them, and I was happy to be home. As I walked into the showroom, I stopped dead in my tracks. Sitting there with a huge red ribbon tied around its midsection was every young budding fighter pilot's dream machine. Its white convertible top was down. It looked so low and sleek sitting there. It was blue with a white stripe down the middle running from the back edge of the rear fender all the way to the front where it narrowed down into an arrowhead shape. Located just above the stripe on the front fender were the magic words in chrome "TRI-POWER." I walked slowly up to the Pontiac Chieftain and just shook my head. I had never seen such a beautiful car. I leaned over the left front door and took in all the dash and upholstery. My parents came walking up to me smiling and said welcome home. I hugged and kissed my mother and shook my dad's hand. (This was the 1950s. No dad hugging allowed.)

I couldn't take my eyes off of the convertible. My dad said, "Want to look under the hood?" He had a big grin on his face as he popped the hood open. I quickly joined him in leaning on the front fender and peering into the engine compartment: "389 Cubic Inches" read the decal on the air cleaner. I think my little Ford coupe had around 100 Cubes. There, perched on the top of the intake manifold, were three carburetors. I had been reading about this engine. It was considered to be in the Super-Stock class and was about the hottest thing coming out of Detroit. I was drooling all over the spotless fender.

I stood up and just shook my head and started thinking about how I could manage to get one some day. I asked, "Who in the world does this car belong to?" My mom just smiled at me in her quiet way, stuck her hand out, and handed me the keys!

I honestly don't remember anything about what happened next. I don't even remember driving that wonderful machine for the first time. I must have been in some form of mild shock. It was a total, unbelievable surprise. I don't have any memory of the rest of that day, or for that matter, the rest of that Christmas leave.

Everyone should be lucky enough to have parents like I do.

The Honor Code, *again*

Hondo Air Base, Texas, January 1958

My shiny new Pontiac convertible was parked in the lot provided for Aviation Cadets. We were not allowed to drive our cars except when we were on our somewhat rare weekend passes. One day shortly after returning from my Christmas leave, I was ordered to report to my Tactical Officer. He was a Lieutenant who was responsible for overseeing my particular group of cadets in military matters.

I reported to him in the usual military manner. He seemed unfriendly and slightly hostile. I couldn't imagine what I might have done to warrant this sort of attitude on his part. I couldn't think of anything. I still hadn't even had to walk a single tour yet in my short career as an Aviation Cadet. It was also very unusual to be called in to see a TAC officer. Usually most matters were handled by a ranking cadet member of our upper class.

He didn't ask me to sit down. He didn't even give me a "parade rest" order. He left me standing there in front of his desk at a stiff attention. "What is this all about?" I was asking myself. I didn't dare ask him with his obvious hostile attitude toward me. I just stood there silently while he glared at me.

"Don't you respect our Honor Code here Mr. Cook?" he finally asked.

"Yes sir. Of course I do." I replied.

"Then why have you chosen to violate it in such an obvious manner?" he replied.

"Sir, I don't know what you're referring to, Sir." I nervously said.

"Then maybe you're too ignorant to be here in pilot training Mr. Cook." he said.

"Sir, one thing is for certain. I'm ignorant of what you are referring to as an Honor Code violation, Sir." I stated. (My nervousness was rapidly being overcome by my anger, but I couldn't afford to show it. I struggled to maintain an appearance of humility.)

"You went home on leave and returned with a new convertible, did you not?" he asked.

"Yes sir, I did," I answered.

"You are *not* your own boss here Cadet! You must ask my permission to buy a new car or even a used car for that matter," he boomed out.

"I did not buy that car Lieutenant. My parents gave it to me for a Christmas gift," I stated firmly.

"I do not believe you, Mister Cook," he said coldly. "And lying is another Honor Code violation!"

My blood ran cold at being called a liar by this little wimp sitting so smugly there before me. What a situation. I wanted to whip his butt so bad for these totally false accusations, but I had to stand there and take it. (I'm sure some psychologist would say it was good for my intestinal fortitude or something, but it sure didn't feel very good at the time.)

"Sir, I am not lying, and I resent being called a liar. I'm also very aware that no matter what action I would like to take at this moment, I must stand here in this very unfair situation and take this 'crap' from you, SIR!" I went very heavy on the last sir.

He sat there for a long time without moving. I could tell that he was staring at me although I did not look at him. I was shaking from my anger, and I know my face was red.

"I'm going to contact your parents Mr. Cook. You had better be telling me the truth. And even if it is true, you should have gotten my permission to accept the car."

I had never heard anything so ridiculous in my young life. This guy was jealous! Some little low-class aviation cadet had a better car than his, and he couldn't stand it!

"Lieutenant, you feel free to call them. Please talk to my father. I'm sure he'll be interested to hear what you have to say. I wish he was here right now so he could have heard this whole conversation," I replied as I glared back at him.

"Cook, you're dismissed!" It was obvious that he was very angry. I saluted and did an about face and got out of that First Lieutenant's office before I said something else. I never heard anything more from him. That includes an apology.

As my dad always said, "That's why there are more asses than horses!"

Descend lower

Hondo Air Base, Texas, 1957

Recalling my night time incident at Hondo, it certainly wasn't funny to me; however, I recall an incident at night which happened to another student that I thought was hilarious, particularly since no damage was done and no one was injured.

The call to the control tower sounded amazingly calm, considering. It was a black night, the student pilot was solo, and he had an "unsafe" main landing gear indication on one side.

The control tower asked his position.

The answer was "South field." It was a little hard to understand because the student pilot was one of our South Vietnamese aviation cadets. The tower told him to bring it by the tower and they would try to see if the gear looked down in the lights from the ramp and from his own landing lights.

Several minutes went by and the tower operators were straining their eyes trying to pick up the T-28 in the darkness. Finally they asked, "T-28 with the gear problem, say your position."

The answer, "North tower."

"Well, we missed you. Bring it by again," said the tower.

The answer, "Roger," with a heavy Vietnamese accent.

A few minutes later, following another unsuccessful attempt, the tower suggested to the student pilot that he should descend lower so they could see him. The calm student pilot replied that he could not comply with their suggestion. When the tower asked why, he answered that he was already on the ground! Sure enough, they looked down and saw a lone T-28 taxiing back and forth out on the ramp in front of the tower on the unsafe landing gear.

Fortunately they got him stopped before the gear could collapse. It seems the student pilot had called the tower on the airborne frequency instead of the ground control frequency. He had made the first call sitting in his parking spot south of the tower. When they asked him to bring it by the tower, he complied by taxiing about a quarter of a mile from the parking area!

When the student called the tower on their airborne frequency, they made a logical, although totally incorrect assumption. We all know what assuming does to us sometimes, don't we?

By the way, Aviation Cadet Thai De is said to have flown gallantly as a fighter pilot for his country long before the rest of us entered the "conflict." I hope he is alive and well and taxiing along on all safe gear today!

Aaa-haaa, San Antoine!

Hondo Air Base, Texas, spring 1958

It was a beautiful warm day in south Texas. Even better, it was a Saturday and we had a weekend pass. And I had my new Pontiac convertible. I and two of my cadet buddies had the top down, our Levis (with no belts, of course), our shirt sleeves rolled up, our collars turned up and were wearing our official U.S. Air Force pilot sunglasses. We were just about as cute as it got in 1958.

We had a plan. We were going to the big city, cruise around with the top down and flirt with girls. First, for kicks however, we were going to go out to Lackland Air Force Base in the morning and be there when the buses arrived with the newest aviation cadet class. One of the guys at Hondo was from the Air National Guard, and he had a friend who was in the new class. That's how we knew when they would arrive.

We were parked down the street from the cadet barracks when the buses drove up. At first we were just going to watch the fun as the new cadets got their rude welcoming to the Aviation Cadet Corps. Then someone got the bright idea that we should join the new guys and pretend to be some of them. After all, they would all be in civilian clothes and so were we. Since we were so superior now and were in the upper, upper, upper class, we could have some fun driving the TIs wild and then just walk away while they were still chewing us out, making them even wilder. Good idea!

Someone else has always thought of it first. Evidently our idea was far from being original. The TIs perhaps had a lookout but more than likely it was our official U.S. Air Force pilot sunglasses. Dumb! Not one of us thought of that dead giveaway. One of the senior TIs even thought he knew *who* we were, got in our faces anyway and impolitely invited us to leave. We left!

Oh well. Chasing girls is a lot more fun anyway. We thought.

We were so cool. We just knew that everyone was looking at us, especially the girls. I was the coolest though. It was my car!

We had spotted this one particular vehicle. It became our primary target. It was a 1956 Chevrolet Bel Aire four-door, brown and white. Three members of the fairer sex were sitting in the Chevy giggling. They had been trailing back about a car length in the left lane. I had slowed down several times to try to get them to come up beside us, but they kept hanging back. Finally we stopped at a red light, and they decided to make their approach. I was slouched as coolly as possible with my left arm hanging over the driver's door and my right hand on top of the windshield.

I turned my head and smiled at the girls in the car. The one in the right front smiled back and said, "Nice car."

I grinned a little bigger and said something super cool like, "Thanks."

She then smiled sweetly (you know how the fairer sex can do that) and said, "You sure are cute!"

Of course I agreed with her completely and said "Well thanks." I was really on a roll now. I could tell that the other two guys in my car were in awe.

Then she laughed and said, "But I already have a dog!" The three girls roared off in their dad's Chevrolet, laughing all the way. I let them.

As we were eating our Texas size T-bone steaks later (just the three of us), I was hoping the other two guys would choke because they were still laughing so hard.

But they didn't.

Number one?

Hondo Air Base, Texas, 1958

Sometimes I think there's a heavenly being somewhere who keeps track of our mindset. Then, when we need an attitude adjustment, the being instigates that adjustment before we get too cocky. Looking back, I think what happened was funny and I know it was good for me, but I sure was upset at the time.

The next to the last night of our training at Hondo, we held a banquet. I had heard that awards were always given to the top student in each division in our training. Those three divisions were Officer Training, Academic Training, and Flight Training. We were graded on everything and evaluated at the end to determine our overall class standing. That was very important to us, as our next assignment depended on it. If you were high enough in your class, you got jets. Everyone else went to multi-engine training in the old B-25 Mitchell bomber. After dinner, an officer stood up and announced that he was going to present some awards. I had done very well in officer training, was about sixth or so in academics, but in flying I was pretty sure that I had finished number one. The other two categories were important of course, but this was pilot training. That meant "flyer," and I was probably number one!

As the top student in officer training was announced and went forward to get his award, I began to look both ways to see which aisle was closer. The first winner came back and sat down. He had been given a beautiful pen and pencil set engraved for the occasion. The winner of the academics award was going forward to receive his prize. I was shifting in my seat knowing that I would be called next. In fact, some of the other student pilots sitting around me were already congratulating me as they too thought that I had finished first. The presenting officer was starting to say something else. I shifted my weight forward in my chair to stand up. I had decided to go to the aisle to my right. *Wrong!* The officer was offering further congratulations to the *two* winners, and that was the end of the presentations. Several of those around me looked at me, and I tried to pretend that I wasn't really half standing, but was merely adjusting my chair. I don't think that my face has ever been as red as it was that night.

I couldn't believe that they weren't giving an award for the top student in flight training. They didn't even mention flight training! I don't know how the angel in charge of such things pulled it off, but I sure wasn't cocky anymore, at least not for a few hours anyway. All I could ever figure out was that the flight training instructors must have been a bunch of cheapskates.

Hondo Air Base was fast disappearing in the rearview mirror of my hot new Pontiac Tri-Power convertible, the wonderful Christmas present from my parents. There were a lot of definite advantages to being the only child of a new car dealer! I had just passed that sign reading "Hondo, Texas. This is God's country, don't drive through it like Hell!" I assumed that didn't apply to driving out of town, as I accelerated behind the 389 cubic inch V-8 with its three carburetors. With the top down, I achieved just below the speed of hair loss and began thinking about the last few days of the Primary Flight training I had just finished. I couldn't say that I was sorry to be leaving. Between my almost busting my own butt on that night flight with the rough-running engine and the chief check pilot almost busting it for me on that instrument flight, I was ready to get the hell out of Dodge, or Hondo. Whatever!

However, I sure could have used that pen and pencil set!

Chase me back to town

Kingfisher, Oklahoma, late spring 1958

I was on my way to jet training at Greenville, Air Force Base, Mississippi. I had detoured to Kingfisher for a few days with the extra time they had allowed us between Primary and Basic Pilot Training. I found out that my reputation, or at least that of my car, had preceded me.

Everyone in the local area who thought they had a fast car had tried to goad me into a drag race and some of them had succeeded. None of them had won. In fact, the boyfriend of a girl that I knew from high school had dropped second gear out of his 1957 Bel Aire hardtop just the night before trying to beat me. Now the girl was mad at me because she was having to walk for a few days.

My car was up on the hydraulic lift in my dad's dealership getting one of its manufacturer's checkups. A guy walked up to me who I had never seen before. He had a badge and was wearing a gun on his hip. He pushed his western hat back on his head and said, "Are you Jerry Cook?" As you can probably imagine, I was a little hesitant to admit it. But I knew he would find out anyway so I told him, "Yes, I am." He stuck out his hand and I relaxed a little as he introduced himself. He was a deputy sheriff. He told me that he had bought a Buick Century hardtop from my dad several months ago and that it was very fast. In fact he said it would outrun my Pontiac any day. I tried not to take too much offense at that. After all, he was packing a gun!

"What makes you think that your Buick will outrun my Pontiac?" I inquired.

"Because nothin' round here has outrun me yet," he bragged!

"Actually, if I didn't think I'd get in trouble with you, I'd say the same thing," I said and grinned.

"I already know that. Word gets around. But you ain't run me yet," he said.

"Well, I guess that there's no way that we're gonna know," I replied.

"Sure there is, that's why I'm here," the deputy stated. I was listening.

"We can go out east of town on Highway 33. We'll start from a dead stop and run for a couple of miles. I'll have my red lights on and if anyone sees us, they'll think that I'm chasing you," he continued.

Boy, that sounded like a setup if I ever had heard one. "How do I know that this is on the level and you won't just arrest me for speeding?" I asked.

"You'll just have to trust me I guess." He smiled as he said it. "After all, I'm a cop ain't I?"

I didn't know this guy from Adam but there we were, sitting in our respective cars side by side, stopped on a straight stretch of Highway 33. We could see a mile or so in either direction, and there were no other cars in sight. I was as nervous as a constipated cat.

He dropped his hand as the signal, and we took off. He had been in the left lane when we started, and I saw him cross into the right one in my rear view mirror. His red lights were flashing. I had expected to outrun him, but not nearly as badly as this. At the predetermined spot, I hit the brakes and then pulled off on the shoulder when I was slow enough. He was going so fast when he careened onto the shoulder that I thought he was going to run into the back of my Pontiac. He jumped out of his car and came running toward me. For a second or two it looked like he was going to pull his gun on me, but then I realized he was just holding the holster to keep it from bouncing.

He jerked opened my driver's door and yelled, "Get out!" I did. He then jumped in my car and shut the door. "Now, chase me back to town!"

I could see my little blue Pontiac convertible getting smaller in the distance and I could hear the three carburetors sucking in air as he continued accelerating. (I had it clocked about six months later at one hundred and forty seven miles per hour.)

I trotted back to the cop's car and turned it around and headed back toward Kingfisher. My Pontiac was long out of sight. As I turned the corner off Highway 33 toward my dad and mom's dealership, I could see my car sitting there on the east side of the building. I parked beside it and got out. The Pontiac was popping as the engine cooled down. I looked around but didn't see the deputy. One of Dad's mechanics named Roy was standing there wiping grease off his hands. He grinned and nodded toward my dad's office. I thought, "Oh man. I'm in trouble now."

My dad and the deputy sheriff were in a deep discussion. I walked back into the shop area and bought a Coke. After about thirty minutes I went back up toward the showroom. The deputy was gone.

My dad walked over and said, "Do you know what just happened?"

I swallowed hard. "Not really," I replied, trying not to give out too much information.

"Well, you just sold a new Black Pontiac Tri-Power hardtop," he said. "I just ordered it."

He hasn't let on to this day whether he knows how!

I never saw that deputy again, but one thing is for sure. I wouldn't have tried to outrun him if I had.

Jets

Greenville AFB, Mississippi, 1958

There it was, the "Mighty Miss." I was crossing over the long bridge into Greenville, Mississippi. I had been driving all day from my parent's home in Kingfisher, Oklahoma. After enjoying a few days of leave visiting with them and some of my hometown friends after Primary Pilot Training, I was reporting to Greenville AFB, for Basic Pilot Training. My class standing had been high enough, and I was really excited at the prospect of flying jets. Only nine months before, I had boarded that Central Airlines DC-3 for my first airplane ride and here I was, about to check out in jets. Can you imagine anything better than being twenty years old, driving a new convertible with three carburetors, and getting to fly jets? I couldn't. Even without the pen and pencil set!

Greenville wasn't a very big town, and I soon found a place to grab a burger and fries. I knew that it would probably be my last civilian meal for a while because I was going to be in the underclass again! As the newest cadets at the base, I knew that we would be restricted to the base for a few weeks. While driving out the two-lane blacktop road toward the base north of Greenville, I was thinking about being the underclass once more, for the third time in ten months. It seemed just when you got to be the upper class, with its ensuing privileges, it was over, and you were under again. That was really getting old.

I was considering turning around and going back to town for a few more hours. I only had to sign-in sometime before midnight, and that was several hours away; however, I wanted to be a little early to try to get my bearings. Besides, I wanted to see the jets. Also, I was really tired and needed to get some rest. I signed in and was given directions to my barracks and assigned a room number. Greenville was a relatively small base, and I found my building and room quickly.

Just a few cadets went to Greenville from Hondo. Others had been assigned to other jet training bases or sent to multi-engine schools. All the other members of my class here were reporting from other primary schools similar to Hondo. I recognized a few faces I had seen at Lackland, but remember, I had been staring straight ahead at attention most of the time. Consequently, the vast majority were new to me. My roommate was one of these. He had attended Primary in Florida and was from Ohio. After a mutual introduction, I promised not to call him a Yankee in exchange for him not calling me an Okie. It only took me a couple of days to begin to understand what he was saying. I think my prior "training" with my Lackland roommate, who was from Brooklyn, somewhat prepared me. I never did understand what he was saying! I guess maybe that's the difference between a Yankee and a damn Yankee.

After unloading my car and putting away my things, I made my bed and laid down on top of the covers. The next thing I knew, my new roommate was shaking me and saying "Chow time." I understood that perfectly, and we went to the dining hall.

No one hassled us until the next morning at reveille. Then the hazing started again. At least there were no TIs, just the upper class. The harassment was much less intense than either Hondo or Lackland. I think we were getting a little more respect because we had progressed this far, plus the upperclassmen were so busy themselves. My life was personally much improved because my nemesis, the upper class cadet who had tigered my bed at Lackland, had been sent to some other base. In any case, except for the marching and calisthenics, it really wasn't so bad.

Greenville possessed two parallel runways that supported a lot of flying activity. This time however, unlike Primary, only a few days passed before we began flying. I met my instructor the first day at the flightline. He was a relatively new second lieutenant. I think that he had instructed only one class prior to us. He was a bachelor and a neat guy. He seemed more like one of us than an instructor. He had been a cadet, so he understood what we went through. I think that really helped a lot. His callsign was "Dirty Boy." We didn't socialize at all with the instructors, but I think his callsign may have been derived from his associations with the ladies. I never did know for sure, and I never asked.

Dollar-ride time again. I had been through ejection-seat training a couple of days before, and I was strapping into the cockpit of the T-33 jet trainer very carefully. They had shown us pictures of the results of accidentally ejecting on the ground. I felt somewhat like I was sitting in a chair with a bomb under it that would explode if I moved around too much. I knew the ejection seat safety pin was installed, but I still was being very careful. Dirty Boy was standing beside me on the ladder, which fastened over the front canopy rail. He gave my oxygen mask and seat connections one last look, tapped me on my helmet, and began strapping into the back seat.

I was inhaling as hard as I could through the oxygen mask. In the T-34 and T-28, we hadn't worn a helmet or oxygen mask. We hadn't flown high enough to need oxygen and had utilized just a baseball cap and headset. I had only worn an oxygen mask for a short time during our altitude-chamber training. I felt the same way then. I couldn't seem to get enough air. I knew I was, but I was still pulling hard with my lungs. I sure hoped that I would get used to it, like they said I would.

"Can you hear me okay, Cook?" my instructor was saying.

"Loud and clear," I gasped through the mask microphone.

"Just relax your breathing, you're getting plenty of air. That's a normal feeling the first few times," he said.

I thought to myself, "I sure hope so." I had never dreamed that one of the hardest things about flying a jet would be breathing, and we hadn't even started the engine yet!

Dirty Boy talked me through the first jet-engine start of my life. It was not as simple as some of the jets that I later flew. The ignition wasn't introduced automatically before the fuel. Sounds like a simple thing, ignition, *then* fuel. But nothing seemed simple after struggling for breath for five or ten minutes! They had said that when you got it backwards, the resulting rumbling noise could sometimes be heard all over the flight line. They had also mentioned that it wasn't especially good for the engine and tended to upset the crew chiefs. "Ignition, then fuel," I said to myself as I waited for the proper engine revolutions. I felt more relaxed as the engine started smoothly and all the engine gauges moved into the proper green zones. We slowly progressed through the checklist items. We were proceeding toward one of the hardest parts of flying a T-33 for a new student, besides breathing, that is!

Taxiing out of the parking area and to the runway seems like a relatively simple matter to the uninitiated. In the T-33, it could be a nightmare. They had designed the Lockheed Shooting Star without a nose wheel steering system. You had to use the brakes to turn. If you tried to turn too sharply, the nose wheel would swivel too far and cock. Once it had cocked, the procedure to get it out was difficult. You had to add power and "jump" hard on the opposite brake. If you just pushed it steadily, it just cocked tighter, and you went round and round in tight little circles, blowing away everything behind you.

Almost every day you would see some poor solo student going round in circles, blocking all the other airplanes behind him on the taxiway. More than once while I was at Greenville I saw an instructor unstrap from a nearby T-33, go jump up on the wing of the circling culprit and uncock the nosewheel. Then he'd crawl back out, slide off the wing, and send him on his way. Then he had to crawl back in his own jet with his student and proceed with his mission. Thank God that never happened to me, but that first day, learning how to uncock it was really difficult. It felt like you were abusing the airplane and that you were going to break something when you did it correctly. I finally got the hang of it, and off we went to the runway.

The old T-33 wasn't exactly overpowered, but in comparison with what I had been flying, it seemed like a rocket. There's a saying about "not getting hurt if your jet crashes on your first flight because you're so far behind it." I didn't believe the "not getting hurt," but I could verify the "being so far behind it" part! We were climbing through five thousand feet and my mind was still somewhere on the runway or thereabouts.

It was incredibly smooth. The engine noise and vibration were gone. Instead of being in front, the engine was behind us and so was most of the noise. I was still struggling to get more air. In retrospect, I realize most of my respiration problem was caused by tension and nervousness. My mind was furiously trying to catch up with the airplane. My instructor was pointing out features in the local flying area, that I couldn't seem to see. I was trying to check instruments in the cockpit that I couldn't seem to find.

On top of all that, I was starting to feel sick at my stomach! I hadn't even come close to getting sick in an airplane before, and here I was fighting to keep down my breakfast. Unlike my dollar ride in the T-34, I couldn't wait for this one to end. I felt totally humiliated. Not like I would have if I had tossed my cookies, but I was certainly upset with myself.

Finally it was over. I was drenched with perspiration although the weather was cool. The fresh air flowing in the cockpit when we opened the canopy felt like paradise. Finally my instructor said, "Let's take our masks off while we taxi in." I yanked off the mask and took my first deep breath in more than an hour. I couldn't believe how good it felt. All my sick symptoms disappeared, and I was so grateful that I hadn't thrown up. In fact, not throwing up was something of a small victory. Several of my classmates had not been as fortunate as I, and they felt ashamed. Some of them became airsick during several flights before their bodies acclimated. I know that at least one member in the other half of our class never did get over it and finally washed out. I thought that was a terrible waste because he had been a good pilot otherwise.

My second flight and all those that followed, went well. I never felt ill again. I also never had any more problems breathing through my mask. I was catching up to the airplane rapidly, and it was becoming more and more fun. I loved the aerobatics. The T-33 was fairly light on the controls and responded well. I began to feel more and more like a fighter pilot. In fact, the T-33 was a derivative of a fighter used in Korea, the P-80. I really enjoyed the flying.

Timing is everything

Greenville AFB, Mississippi, 1958

I can remember being apprehensive about only one thing in Basic Pilot Training.

Dirty Boy loved to demonstrate what he called "diving off altitude on final without gaining airspeed." Translated, that meant being on final approach for landing with the landing gear and flaps lowered and finding yourself too high on the glidepath. He would "dump" the nose sharply and point it at the ground. After losing several hundred feet, he would snatch the nose back up, having only gained about five knots of airspeed. Then he would land. The first time he did that with me, I just knew that we were going to hit the ground short of the runway. We didn't then or any other time that he demonstrated the technique to me. When I would practice it at his prompt, I didn't lower the nose as abruptly or quite as far. It still got the desired result, but he didn't think it was nearly as much fun. I must say that I was glad to finish the transition phase of my training and leave that maneuver behind.

(A sequel to this particular part of my story.)

In the very next class that my instructor trained after I graduated, the following happened. While demonstrating this exact same maneuver to a new student, my ex-instructor misjudged and the T-33 struck the ground before reaching the runway. My source said that one of the T-Bird's main landing gear was torn off. Dirty Boy proved that he was an excellent pilot, considering he got the airplane airborne again and flew it back to Greenville. (The accident had occurred at a small auxiliary field nearby.)

He landed successfully, and no one was injured. I understand that the Flying Safety Officer, who was observing the emergency landing, was very perturbed however. It seems that the canopy, which was jettisoned as part of the procedure, landed on top of his official vehicle and accomplished a substantial amount of peripheral damage. In fact, the whole thing, I believe, was reported in the Air Force Flying Safety Magazine. I was flying F-86 Sabres in Arizona at the time and I remember thanking my lucky stars that I had graduated without that particular experience.

(A much sadder sequel.)

Dirty Boy was killed about six years later in Vietnam. He was flying a B-57 at night and crashed near Bien Hoa, South Vietnam. He was an important part of my life. I learned much from him.

Another incident comes to mind that nearly touched me at Greenville. It very aptly demonstrates the luck of the draw or fate, if you will.

I was scheduled to fly two solo missions one hot August day. I had been assigned one aircraft to fly in the morning and another aircraft to fly later that day. When I returned from my first flight, another student was waiting for me near my parking space. When I shut down the engine, he came up the ladder as I removed my helmet. He said that he had just flown the aircraft that I was supposed to fly later. He was also supposed to fly again, in the aircraft in which I was sitting. He wondered if I might like to trade so neither of us would have to lug our heavy parachutes to the other aircraft in the oppressive heat. That is, if he could arrange it.

I said "Sure, that sounds like a good idea to me."

He thanked me and said that he would try to get it approved. I left my parachute and helmet in the airplane and walked into the flight shack to relax and have a Coke before my next flight. I checked the schedule board and saw that the aircraft change had been approved. The other student's takeoff time was several minutes before mine, and I saw him leave to go to the airplane. I drank my Coke and rested for a few minutes while my instructor briefed me on the things to practice on my next solo flight.

I was waiting in line for takeoff. A four-ship formation was just beginning its takeoff roll. Fifteen seconds after the first two T-Birds rolled, the second two released their brakes and began rolling together down the runway. The first two were beginning a right turn just as the number three aircraft began his liftoff. I saw that number four was trailing slightly. He rotated, but his aircraft didn't lift off. He was on the asphalt overrun, just beyond the runway, when he yanked the airplane into the air. He raised the landing gear, but the aircraft didn't climb. Dirt was being blown everywhere by his exhaust, which looked to be just a foot or so above the ground.

Beyond the south end of Greenville's runways was a row of trees. I didn't think the T-33 would be able to clear them. Just before reaching the trees, the pilot pulled the nose up and skimmed nose-high just above the tops and disappeared. I expected to see a fireball, but could see nothing. Finally, the airplane appeared again a mile or so beyond the trees in a slight climb. One of the other T-33s broke from the formation and joined on the stricken aircraft's wing. As we waited on the ground, the pilot of the emergency aircraft jettisoned the fuel tip tanks and landed. When he passed by in front of me, I recognized the tail number. It was the aircraft that I was supposed to have flown.

I talked to the student pilot later. His engine had thrown several turbine blades that had broken just before he rotated. With only partial power remaining, he had just managed to get over the trees. After clearing them, he had to descend very close to the ground again while gaining enough airspeed to climb. He had been very fortunate, in addition to doing a great job of flying. I asked him if he wanted to trade any more airplanes with me.

He said that although he *wasn't* superstitious, "It would be a cold day!"

Mississippi Luftwaffe

Greenville AFB, Mississippi, 1958

I was waiting at the number one position for takeoff on runway 17 right. I was "holding short" as it was called, when I looked toward the north and saw the four-ship of T-33s descending toward the runway in tight, in-trail formation. I was confused. I had never seen jet trainers so low in formation. Strange, I hadn't heard anything about an airshow, and besides, there was traffic in the pattern on both parallel runways.

As the formation approached closer and closer, the lead T-bird leveled off at about three hundred feet and the other three were bouncing badly, especially number four. They were tucked tightly in trail with the leader, each slightly below the aircraft in front of them and only one or two plane lengths behind. As they passed in front of me, I could see that the leader had two pilots, meaning the instructor was aboard it. The other three planes were being flown by solo student pilots. The mobile control officer began going crazy on the radio. He was yelling "Go around, go around," and shooting red flares. As they swept low over the mobile van, the lead plane pulled up steeply into a closed downwind pattern. The other three attempted to follow. They managed, somehow, to miss each other, but looked like a flock of ducks that had just been shot at by hunters. One by one, they began to pull up out of the downwind as the mobile control officer finally got their attention and sent them out of the traffic pattern.

Finally, I was cleared for takeoff, and all during the flight, I tried to decipher what I had just seen. When I returned, there was a special rebriefing on formation visual signals. It was being presented by the three German Air Force student pilots who had mistaken a "complete electrical failure" signal by the instructor for an "in-trail" formation signal.

In a T-33, a complete electrical failure meant there was only the fuselage-tank fuel available, about 90 gallons. That dictated an immediate return to base and a three-hundred-foot flyby over mobile to signal the failure. Mobile would then clear the aircraft to land by using a large green light pointed at it. When the lead aircraft containing the instructor had experienced an actual electrical failure while instructing his four students in the formation flying area, he immediately gave them the appropriate visual hand signals. They were supposed to have departed the formation and returned as individual aircraft back to Greenville AFB and land. When they had disappeared behind him, the instructor assumed they were doing just that. Having problems of his own to remedy, low fuel and no radio, he did what he was supposed to do. Little did he know that he had three inexperienced student pilots parroting his every move.

Fortunately, the mobile officer figured it out immediately and finally got through to the confused students as to where they were. For those readers who have never done it, formation flying is extremely demanding of your undivided attention. This is multiplied many times over when flying close in-trail. Your aircraft's nose is almost directly beneath the tail of the plane in front. To look away for even an instant could mean a midair collision and sudden death for you and perhaps for the other members of your flight. As you can see, these students' actions, instead of being stupid, speak quite highly for their concentration and flying skills. The only real mistake that they had made was totally misunderstanding the visual signals their instructor had given them. Granted, that could have been disastrous, but it fortunately turned out hilarious, educational, and entertaining.

By the way, they all graduated and went home to Germany to fly F-104 Starfighters. Perhaps you had better "Check your six," which is a fighter-pilot expression for looking behind you. These guys were good, even then!

What about my eyes?

Greenville AFB, Mississippi, 14 October 1958

It had seemed like years, but it had only been fifteen-and-one-half months since I had stepped onto that DC-3 for my first airplane ride. Now here I was standing in the base theater with my mother pinning on my Air Force Pilot wings. I was one proud twenty-year-old. I had also just been presented with my officer's commission after a swearing-in ceremony and was a brand-new second lieutenant. I was elated about all this, but the most exciting thing to me was the assignment I had received. I was to be a fighter pilot!

Williams AFB, Arizona, was to be my assignment for the next six months. There I would attend F-86F Fighter-Gunnery School. After that, the normal progression was to Advanced Fighter-Gunnery at Nellis AFB, Nevada, in F-100s. But first, I had to alleviate a worry that I had been carrying with me since January 1957. I called the flight surgeon's office and asked to speak to him. When he answered, I nervously asked him the following question, "How can you get into pilot training without twenty-twenty vision?"

He answered "You can't."

I said, "I know someone who got in with twenty-ten."

He started laughing. Finally he said, "Twenty-twenty vision is minimum acuity for acceptance. Twenty-ten is even better. In fact, it's twice as good."

Without identifying myself, I thanked him and hung up. I couldn't believe that I had been so naive. I had been sweating my vision for twenty-one months, and it was even better than required. I could at least have checked it out in the library or with a civilian doctor. I think I just didn't want to know. I felt both stupid and relieved. I also knew now why I could see airplanes farther away than most of my friends or instructors and do it with bad eyes! Oh well. I never said you had to be real smart to be a fighter pilot. In fact, if you were, you probably wouldn't do it!

I can't begin to express my excitement at the thought of finally becoming what I had wanted to be since I was five years old. I don't think I realized at the time how fortunate I was. I believe most of us have dreams of what we would like to be when we grow up, but I suspect a very minute percentage of us ever get to do it.

Of course there are those who think that fighter pilots aren't grown-ups, but that's another story.

Sabre

Williams AFB, Arizona, 1958

The major entered the briefing room carrying a cardboard box. He sat it down on the podium in front of him and said, "Be seated, tigers." We relaxed from our position of attention and sat down grinning. We'd never been called tigers before and I knew that I was in the right place. Suddenly he started throwing checklists for the F-86 around the room. One by one, we caught a checklist as it came at us and began to devour the contents.

The F-86 had been the winner of the air war in Korea with a 13-to-1 kill ratio over the enemy MiGs. That had been six years ago, and there were faster fighters being produced, but the F-86 Sabre was considered a sports model, and everyone wanted to fly it. It was the jet equivalent of the P-51 Mustang, which was every fighter pilot's dream propeller-driven fighter. Furthermore, we all expected to leave here in six months bound for the faster F-100 Super Sabre at Nellis AFB.

When he had finished tossing out the F-86 checklists, our flight commander introduced himself. I can't remember his name, but I do remember that he was an ace in Korea. In fact, I found out that nearly all of our instructors had flown in Korea, and several of them were aces. I felt like I was joining the "first team." I remember the next thing he said very clearly.

"It's apparent that you all are excellent pilots or you wouldn't be here. There's lots of information in that checklist that I gave you, and it would behoove you to learn it all; however, I know you won't. But the best pilot in the world can't fly an F-86 very far with the engine flamed out, so at least learn the 'Engine Airstart' and the 'Ejection' procedures. The ejection procedure has an extremely important step that this slide will help you remember," he said.

He then showed us a picture of a helmet that was worn by a pilot who had forgotten the step. The top half of the helmet was missing. He then informed us that the top half of the pilot's head had been missing also. The canopy on the F-86F model that we would be flying jettisoned rearward instead of upward like most other jets. As that occurred, it would strike you just above the eyebrows if you were sitting up straight. Of course, the procedure called for you to duck forward in the cockpit as you jettisoned the canopy during the ejection sequence. That presentation was so potent that I think I spent my first two or three flights saying "Don't forget to duck" to myself.

Me first. I wanted to be the first one in my class to fly the Sabre, but someone else beat me to it. After what happened to him, I was glad. We were sitting in the briefing room studying and waiting our turn. We were also waiting for the lucky guy who was out getting to be the first. As he strutted into the room with his G-suit on and a big smile, we all surrounded him and began quizzing him on his inaugural flight. He had us going pretty good about how easy it was for him and how calm and cool he had been; then the Master Sergeant walked in. As he spotted the fledgling fighter pilot, he asked him if he would please accompany him back to the flight line.

When the new, cool, fighter pilot asked him why, he replied, "So you can shut the damned engine down before you leave the airplane this time!"

Seems he had been so cool, he had just parked the airplane, got out of it with the engine still running, and came inside to brag.

I guess sometimes it's better not to be first.

My turn. The F-86 was a single-seat fighter. There were no dual or two-seat versions of it. There were no simulators. The first time you flew this plane, it was all yours. Your little pink butt was the only one strapped in there. There was no one to help you save it but you, which was a rather sobering thought.

My instructor was hanging on to the left side of my F-86. He had his left foot in a kick step on the side and the other on the leading edge of the left wing. He was fastened to my seat belt with a harness so he wouldn't fall off. I started the engine with him watching all my moves and progressed through the checklist to the point just before taxi. He then unbuckled, trotted to his airplane parked next to mine and started it while I finished strapping in. My first flight in a fighter was about to begin. I taxied toward the runway with my instructor's aircraft close beside me and slightly aft. He was going to be there during the whole flight in the "chase" position. It was the next best thing to having him in the airplane with me.

Every noise, every sensation that I would experience in this airplane would be for the first time. It was quite a different experience than I'd ever had before. It felt almost like my senses were being overloaded. I wondered how I was going to know if something was normal or not! All I could do was have complete trust in the fighter pilot on my wing as he talked me through the flight. It was the first lesson in the confidence that a fighter pilot needs to have in the skill and dedication of his fellow fighter pilots. Without it, individual prowess is wasted and it becomes an infeasible mission.

My chase was saying "Ease back on the stick."

As I did, I was saying to myself "Don't forget to duck!" At his prompt, I raised the gear handle, and we were climbing. As I raised the flaps, I remembered that they had said that the aircraft would nose up slightly, unlike the T-33, which would sink a little. Sure enough, the nose rose slightly, and we continued climbing away from the base and turned toward the transition area.

I am disappointed that I can't remember much about that first fighter flight. I do remember that the plane was an absolute dream to fly. It was like a sports car. You just seemed to think about performing a maneuver and it was done. It was light on the controls, and the visibility was outstanding. I remember wondering if all the noises I was hearing for the first time were normal. I remember telling myself "Don't forget to duck" about every thirty seconds until we were back on the ground. Then I replaced that phrase with "Don't forget to shut down the engine."

I was finally a fighter pilot. At least I thought I was. But in retrospect, I know that I wasn't, yet. I was a pilot, in a fighter plane. There are a lot of those types around, and I met several during my career. I became a fighter pilot later—some never do. But that thought aside, just ahead lay some of the most enjoyable flying in my life.

Tiger Four. "Four, I want you joined up in fingertip formation before I complete ninety degrees of turn after takeoff. Three, you had better not let him beat you in. Two, you had better be sitting there on my wing for thirty degrees before three gets in."

Our instructor was briefing us on the takeoff part of our mission to the gunnery range. I loved taking off as number four. I could see everyone else, and I got to start my turn for joinup as soon as I lifted off the runway. If I didn't, I could never get into close formation before the leader turned ninety degrees. It was really fun. On several occasions I had beaten three and joined on his wing before he caught the rest of the flight. I always made that my goal. We took ten-second spacing between aircraft instead of taking off in two-ships. That way we all received practice in joining up, one of the most difficult things to do correctly in formation flying. We would practice wing takeoffs later.

I beat three again. I couldn't believe that I was getting paid for this. After joinup, we would head straight to the gunnery ranges near Ajo, Arizona, or to the air combat area for practice. I can't really remember any individual missions except one at night, which I'll recall later in this "Sabre" section. The rest just seem to blend together in my memory.

One incident which does come to mind though was not one of mine, thank goodness!

First kill. Air-to-air gunnery was one of the more difficult things to accomplish at our level of expertise. A T-33 jet trainer would tow a six-foot-by-thirty-foot banner behind it on a very long cable. As it flew through the firing range, a flight of four F-86s spread out one behind the other would each in turn make repeated firing passes at the target.

Each fighter pilot had to maintain proper spacing behind the fighter in front of him to assure safety. The most dangerous situation for the fighter that you were following was to reverse its "S" turn toward the target too late. That placed it below your field of view. You could then mistakenly assume that another fighter in the flight was the one that you had been behind. As that misidentified fighter ended its firing pass and pulled up from the target, you would then think that it was safe to fire at the target banner. Wrong! There was still the fighter that would soon be flying between you and the target from beneath your field of vision. It was not a healthy place for him to be.

Each F-86 was loaded with fifty-caliber target-practice ammunition. The rounds were marked in different colors for each aircraft and would make a mark on the target banner when passing through it. The marks would be counted later, and your percentage of hits scored would be figured. Of course, they were not explosive like they would be in actual combat, and it's a good thing that they weren't this day.

I remember well the names of the two students on this particular mission, but I won't name them for obvious reasons. I don't remember who was behind whom, but during one of the passes, an aircraft flown by one of them started trailing smoke, and the engine flamed out. The pilot then performed a fantastic feat of flying. He deadsticked the Sabre into a little civilian airport on the edge of the gunnery range. The runway was so short that after the engine was replaced some of the fuel had to be drained to make it light enough so an instructor could fly it out.

Later, the instructor leading the flight became suspicious and checked the gunnery film from the other aircraft. I saw it later, and sure enough, there came a trailing F-86 flying right through the gun camera film. Before he made it to the edge of the film, he was smoking. The engine mechanics also confirmed the shootdown. The young tiger was one-fifth of an Ace.

Of course that's not the percentage or the spelling of the term that his victim used when describing him. He used an "A" and a double "S."

Their instructor deemed that both were at fault. They had to repeat that mission to a more successful conclusion.

Night fright. Just on the south edge of Phoenix, Arizona, lies a high ridge of rocks called South Mountain. At least that's what we called it then. There was a beautiful moon over the "Valley of the Sun." I was completing a night "round-robin" cross-country flight, which meant that I had taken off from and would eventually land and stop at the same base. I had flown a triangular flight path basically around the edges of Arizona and had completed it over Phoenix. The city lights were glowing high into the night sky as I began my descent over the well-lit valley. I was having fun turning and pulling Gs, rolling from side to side looking out for other aircraft as I let down.

The airbase was located southeast of the city, and I began to generally head in that direction. I remember that I was looking toward the south when I suddenly realized that something was terribly wrong. At that instant, the lights to the south disappeared. Thank God that my youthful reaction time overcame my youthful ignorance. As I snapped the wings level and pulled for my life, I could see details of South Mountain in the moonlight through my forward windscreen. Individual rocks and bushes were discernible as my F-86's nose struggled to rise. I had shoved the throttle full forward to attempt to maintain some cornering speed, and the Sabre was trying.

I remember seeing the ground flash by in my peripheral vision out of the sides of the canopy as the fighter shuddered through the bottom of the dive. The F-86 clawed for altitude in the night sky. It seemed as if it wanted to get away from the ground as much as I. Finally, I started breathing again and my heart felt like it had just restarted. As I leveled off and headed back to the airbase, my memory cells were very busy storing what had just happened.

Since that night over thirty-six years ago, I have carried with me a very healthy fear of letting down over unfamiliar terrain at night.

If we are to live, we had better learn if we're given another chance.

Bad news. Quite in contrast to the first day that I had sat in that briefing room five months before, my mood was sad and dejected. My class had just learned that instead of being sent to Nellis AFB for F-100 training and advanced fighter-gunnery school, our destinations were to be drastically different. We were to be sent to either the Basic Instructor School in the T-33, or to co-pilot training in the B-47 bomber. It seems that President Eisenhower had just been convinced by the four-star general in command of the Strategic Air Command that there would never be any more limited wars, only a full-scale World War III; therefore, the Tactical Air Command, with its fighter aircraft, should be decreased in size and the Strategic Air Command, with its long-range bombers, should be increased in size.

There would be no more assignments to the F-100 from Williams AFB. Instead, those of us who were high enough in class standings could have our choice of T-33s in Air Training Command or the B-47. Everyone, who got a choice, went to T-33s! The rest got B-47s. All that hard work and sweating to be a fighter pilot had just been flushed down the toilet by a four-star general, who has fortunately been proven totally wrong. Well, at least my hard work and sweat did allow me to stay out of his command. I was really feeling dejected, but my friends who had to go to the B-47 were devastated.

At least, we still had a few weeks left to fly the F-86.

I wonder where in the world we're goin'?

Vance AFB, Oklahoma, 1959

It was nearly spring, and you know what a young man's "fancy" turns to. I have no idea what a "fancy" is, but it must have something to do with girls. My girlfriend was in Kingfisher, Oklahoma, and I hadn't seen her in a long time. We had dated through most of high school and had become engaged several months prior. She was a fine and sweet young lady, and I was anxious to be with her again.

During F-86 Gunnery School each student had an opportunity to fly a cross-country flight. It was part of the training and just reviewed the air traffic control system and the navigation techniques involved in flying from one point to another. It's different and more involved than just taking off and landing at the same base all the time.

The flight had to be made with an instructor in another aircraft and could involve up to four F-86s including his. The problem was finding somewhere that everyone wants to fly to. I must have been very persuasive in my younger days because I talked two other F-86 students and my instructor into going to Vance AFB at Enid, Oklahoma. Not everyone wants to fly to Enid. It's not exactly a Las Vegas or a Disneyland or some other big entertainment mecca. It's a very nice town, but I still don't know how I convinced them.

The reason I wanted to go there was blonde and blue-eyed. Kingfisher was about forty miles south of Enid on U.S. Highway 81. It was right in the middle of Oklahoma wheat country. (In fact, it's nickname was "The Buckle of the Wheat Belt." I wonder who was responsible for that?)

My parents met me at the Air Force Base when we landed. As I said goodbye to the other pilots, my instructor told me when to meet them two days later for our flight back to Williams AFB, Arizona. We would have to stop somewhere en route for fuel, but he didn't know where yet. I gave him my parent's telephone number in case he needed to reach me, and away we went toward Kingfisher. I had a date that night and was excited to see her. My parents and I had a nice "catch up" conversation, and I received all the latest news about my friends and relatives.

That evening I picked up my girl at her home, and we went to the movies. Then we cruised around town and talked to a lot of our old former classmates in high school. It didn't take long to drive across our town so we "dragged" Main Street, as we used to call it, several times. After that we went to the 81 Cafe on the south edge of town and ordered fries and a Coke. (Ever see "American Graffiti?" The evening was just like that.)

Things sure hadn't changed much. A lot of my friends were still in town working at various jobs while a lot of them were away at college. Of course, I tried to do a little bragging about being a fighter pilot, but no one seemed to care. It's not any fun bragging if no one is impressed.

It was late—about eleven (these were the 1950s). I took my girl home, and we went in and sat for a while and talked. We made plans for another date the next day, and I headed to my parent's house. So went the weekend.

Sunday morning, our planned day of takeoff from Vance, it was raining. The phone rang. It was my instructor calling to say the takeoff would be delayed at least another day. It wasn't because he loved Vance or Enid so much. It was because as student pilots our weather minimums were so high that the weather had to be almost VFR (visual flight rules, usually a one thousand-foot cloud base and three miles visibility). He said that he would call me the next day if the weather got good enough to go and to just stay and enjoy my visit. I can't say that I was upset at the news. I called my girl and we made some revised plans.

I was at her house the next day about noon when her phone rang. It was my mom. She said that the instructor pilot had called and the weather had just gotten up to minimums. He had said to get there as soon as I could, and they would have everything ready to go and my aircraft would be preflighted. All I would have to do was jump in and go.

Away we went. We dropped by my house and picked up my stuff, and I jumped into my flight gear. It was less than a hour later when we drove up by the flightline. There sat our four F-86s all ready to go just like advertised. One of the other students tossed me my helmet and said "Your bird's preflighted. You're number four, on my wing," and strode off toward the jets. I got a swift goodbye kiss from my girl and hurried along behind him. The other two pilots were already climbing into their cockpits. I climbed into mine.

The clouds hadn't gone away. They had just risen enough for us to leave. As I entered them, flying formation on the other student's right wing, I suddenly thought, "Where are we going?" but I couldn't and wouldn't ask. I knew the air traffic controllers would hear me, and I didn't want to sound as dumb as I felt.

All during the flight I stayed close to the right wingtip of number three, my element leader. I didn't want to get separated and have to call Air Traffic Control. I could just hear them, "What is your destination?"

I didn't know the answer to that question until I read the sign in front of the Base Flight Operations Building: WELCOME TO WEBB AIR FORCE BASE, BIG SPRING, TEXAS.

Evidently when a young man's fancy turns to "whatever," his brain turns to mush.

Get your kicks on Route 66

U.S. Route 66, New Mexico, April 1959

I was depressed. I *wasn't* going to Nellis Air Force Base at Las Vegas, Nevada, to fly F-100s. None of us were since General Lemay had convinced President "Ike" Eisenhower that the Tactical Air Command wasn't that important anymore. There was a huge struggle for Air Force funds going on, and the Strategic Air Command was winning. No more F-86 fighter gunnery graduates were "pipelining" to F-100s, the frontline Air Force tactical fighter in the days following Korea. TAC was shrinking. SAC was getting "fatter."

As I mentioned earlier, I was one of the lucky ones who got to make a choice between Air Training Command T-33 Instructor Pilot or Strategic Air Command B-47 co-pilot. It wasn't a difficult choice for me or any of the other frustrated fighter pilots who *had* a choice. We wanted no part of SAC or its bombers.

I had signed out at midnight to get the few extra hours for the long drive to Craig Air Force Base at Selma, Alabama, where the Air Training Command held its Basic Instructor's Course in the T-33. I wanted the extra time to detour and visit my parents and friends at Kingfisher on the way.

I had headed north out of Phoenix and intercepted Route 66. It was the main east-west highway in the 1950s and for many years prior. It basically followed the same route as Interstate 40 does today. It was about four-thirty or five o'clock in the morning, and I had passed through Gallup, New Mexico, a few minutes before. It was really dark, a near moonless night on a black deserted stretch of highway. I remember that I had been running on a section of three-lane, the center lane being for passing. I had left the three-lane and was now on a narrower two-lane section. The speed limit had dropped by ten miles an hour, but I had not. There was no traffic. There wasn't a light to be seen. I was already tired, and I had just gotten started on my long cross-country drive.

The headlights and flashing red lights lit up the inside of my convertible like daytime. This guy was ten feet or so behind my bumper, and I hadn't seen a thing. I recall thinking that I must have been even more tired than I realized. Well he had me. I was speeding by about ten or fifteen miles per hour. There was no way around that. I pulled off on the shoulder of the highway. I rolled down my driver's window and reached for my billfold laying on the seat.

"Let me see your hands," the officer said very loudly.

I held my hands up where he could see them in the cruiser's bright lights.

"Get out of the car," he ordered. I complied. I tried to look at him, but his headlights were still on bright, and all I could see was this dark figure. I couldn't discern any features at all.

"Let me have your driver's license," he continued. I explained that it was in the seat, and he stepped closer and shined his light into my car.

"Okay, get it," he ordered. I complied with that order too. As I did so, he shined his light through the quarter window into the back seat. My clothes were hanging there. My Class A uniform was on the far left side of the rack nearest the window. He paused and looked at it for a second.

"Oh, so you're one of those smart-aleck Air Force pilots," he snarled. "You guys are all alike," he said as if he had a bad taste in his mouth.

I didn't utter a word as I felt I was at a great tactical disadvantage.

"Well smartass, guess what we're gonna do. I'm gonna write you a speeding ticket, and then we're going back to Gallup and you're gonna pay it personally to the Judge."

That really got my attention. Gallup was about thirty miles behind us. That was going to be sixty miles out of the way. I did *not* want to go back to Gallup.

"Can't you just write me a ticket and let me mail the money back," I asked? (How he had gotten up right on my bumper without my seeing him had been puzzling me since I had stopped.)

"No, you Air Force puke pilot, we're going back now," he said raising his voice.

That did it! I raised my voice in return.

"All right Officer. Write your damn ticket, and let's go back to Gallup. I don't know what your problem is. Maybe a pilot stole your wife or girlfriend. Maybe you washed out of pilot training or something, but it has nothing to do with me. I admit that I was speeding. I was driving at least ten or fifteen miles over the limit, but I'll bet the judge will agree with me that chasing me down Route 66 at night with your lights all off is a hell of a lot more dangerous than my speeding. Also I want your name and badge number for my lawyer. We'll see who pays for this chicken-shit ticket!"

I really didn't know what he would do. I had taken a "shot in the dark" about him chasing me without lights.

It must have hit the target. He didn't say anything for a few seconds, but he had stopped writing the ticket. Finally, he straightened up and cleared his throat.

"Lieutenant, maybe I was coming on a bit strong. Guess I'm a little tired and irritable, and you're right. You're not responsible."

I wasn't buying a word of his baloney. I knew it was my remark about the headlights that had caused this sudden humility.

"No. Write the ticket, and let's go to Gallup. I want to talk to the judge face to face." I was bluffing of course.

He suddenly tore the ticket in half. "I apologize lieutenant. Can't we just let this drop?" he asked.

I hesitated as long as I dared and then said, "Okay. I guess no harm's done. We can drop it." He mumbled something about driving safe and walked back to his car.

I stood there and watched him. He dimmed his lights and turned back toward Gallup. I got back in my car and roared back onto Route 66. I was thankful that I wasn't turning back toward Gallup also. As my excitement and anger wore off, I began to realize how very weary I was.

Only seven hundred and fifty miles to go. Driving the speed limit didn't last very long at all. I was tired and I was hungry. Albuquerque was next.

I had lost track of time. I was a "zombie." It felt like I had been driving for days instead of hours. It also seemed such a waste to be driving sixty-five miles an hour. I had discovered not long after getting my Pontiac that it seemed to enter a "groove" at about eighty-five. The engine sounded smooth and effortless at that speed. After all, it was daylight now and there still wasn't much traffic.

I left a small town and began accelerating. A few miles after that I remember passing a sign that read, "Texas State Line, 10 miles." I was at my cruise speed of eighty-five as I crested a small hill. There it was just sitting there with its radar aimed right at me: a jeep with some kind of Sheriff or Highway Patrol markings. I looked farther down the highway, and sure enough, a New Mexico State Highway Patrol car with one trooper in it listening to the radio calls from the jeep and another one was writing a stopped motorist a ticket.

"Texas State Line, 10 Miles," flashed through my mind. I don't know for the life of me why that made any difference, but in my fatigued state of mind it seemed to. Or maybe it was the fact that I had already been pulled over by a New Mexico State Cop a few hours before. Whatever the reason, the next thing I did was one of the dumbest things I have ever done.

I pushed the accelerator to the floor. The cop writing the ticket heard the air sucking into the three carburetors and the exhaust pitch rising. He turned and looked. Then he threw his ticket book in the air and ran back to his big Chevrolet and jerked the door open just as I passed. I was already over a hundred. I remember thinking, "What in the world am I doing?" Then I thought, "It's too late to stop now!" It wasn't of course, but I thought at the time that it was.

"Texas State Line, 10 Miles" read the sign in my memory. The speedometer was pointing well past its maximum reading of 120 when it broke. The Highway Patrol Chevy was doing the best it could, but it wasn't good enough. He was slowly getting smaller and smaller in the rear view mirror.

"Gas!" "Oh Crap," I thought as I glanced down at the gas gauge! It was less than an eighth of a tank, and I imagined that I could see it dropping as I looked. Maybe it wasn't my imagination. At whatever speed I was running, there's no telling how much gas I was burning. Now I was afraid that I was gonna run out of gas at this rate, but I sure could not slow down.

The Patrol car was just a dot in my mirror now. I was passing the few cars ahead of me at probably twice their speed. I saw some of them weave in my mirror after I roared by. It probably scared some of them half to death. I just couldn't believe what I had gotten myself into to. This was really dumb!

There it was. I could see the Texas state line ahead. I looked in my mirror and did not see the cruiser. I was afraid that I would find a Texas trooper waiting for me as I crossed his state line, but there was none. I guessed that they didn't have anyone close enough to get into position in time. Now what?

I took my foot off the gas, and my convertible began to slow. "I have to get off Route 66 for a while," was my next thought. I was back to what I thought was near the speed limit. At least I was maintaining the speed of the other cars. I kept checking my rearview mirror every couple of seconds. My heart rate was probably back down to one thirty or so.

I took the next blacktop road to the right. I hoped there was a gas station soon because I was bumping near empty. Finally a little cafe with some pumps out front appeared, and I stopped.

I bet the people at the cafe wondered why I kept hanging around so long. Finally after about an hour or so, I headed back north to the highway and turned east, driving at the speed limit. I didn't really relax until I crossed the Oklahoma state line.

I don't know what happened to the New Mexico Highway Patrol. I guess I just had too big a head start on them.

Or, maybe they decided that they didn't want anyone that stupid in their state and just let me go!

Laredo By the Sea

Laredo AFB, Texas, 1959

I still haven't fathomed why they referred to my new home that way. There was no sea. There was no sand, unless you wanted to call the dirt that was everywhere sand. It covered most of the streets, even those that were concrete or asphalt. Fortunately, that didn't describe the airbase, too. It was neat and clean and contained all the necessities for a new instructor pilot. That included the Bachelor Officers Quarters where I would live for the next two years.

I had just hung my clothes in the closet and laid down on the bed when I first heard the strange noise. I listened and listened but could not figure it out. It sounded like an animal of some kind, and it seemed like it was in the BOQ. Finally, I got up and went into the hall. It was emitting from somewhere down the hallway. I followed the weird sound. Halfway down the hall was an open door. Inside the room was a young man with a blond crew cut, sitting in his shorts with his feet up on his bed. He had a rifle across his lap and was blowing into some sort of whistle. He would listen for a moment to a recorded noise and then imitate it with his whistle.

When he saw me, he smiled and asked "Want to try it?"

I said "Try what?"

He said "My coyote call."

I asked "Why do you want to call a coyote?"

"Because that's about all there is to do around here when you're not flying. You'll see," he answered.

It seems that the noise was supposed to imitate a wounded rabbit and that would cause the coyotes to come running. The young man's name was Stan, and we became good friends. We happened to be assigned as instructor pilots in the same squadron and flew quite a lot together. Occasionally after that day I would hear the "wounded rabbit" calling down the hallway; however, I guess Stan never did get it quite right because I never did see any coyotes in the building. It did turn out he was right about the lack of things to do in "Laredo By the Sea."

Laredo AFB was a lot like Greenville AFB in its design. It also had two parallel runways about eight thousand feet long. Like Greenville, they were oriented north and south and were far enough apart to support simultaneous flight operations. Air Force pilot training was going strong in 1959, and the base's T-33s flew nearly every day and sometimes far into the night. This south Texas paradise was to be home for the next four years of my young life. It wasn't F-100s, and it wasn't the Tactical Air Command, but at least I was flying.

Anyone who has been an instructor pilot will probably tell you that they learned more about flying as an instructor than they ever did as a student. I know that was certainly true in my case. Here I was a brand-new second-lieutenant pilot with possibly a total of three hundred hours of flying time imparting my "vast" knowledge to my students. Well, at least I knew more than they did, and that's what counted. I figured that if I could just teach them what I knew, they would be okay. At least as okay as I was.

I also was determined to emulate my former instructors. I would try to pass on Dirty Boy's sheer enthusiasm for flying and imitate Mr. Kurkendal's quiet, calm, low-key approach to instructing. I was an extremely proud instructor when my first students graduated. One was second in his class, and the others were in the top ten to fifteen percent.

They didn't know it, but they owed a lot to Mr. Kurkendal.

I'm your instructor. I'm here to learn.

Laredo AFB, Texas, 1960

I say again that I feel my years as an instructor probably taught me more than I ever learned as a student. I think most instructor pilots and former instructor pilots will agree; however, I not only learned from my *students'* mistakes.

I was twenty-two years old. I didn't have that much flying time yet, only about six hundred hours. I wouldn't say that I was exactly an expert in my field, but I was eager and enthusiastic. I *was* also somewhat of a "smartass." I thought that I was a pretty good "stick," even with limited flying hours. I think that some of that attitude was instilled in Fighter Gunnery School, and I still longed to be flying a fighter. Whenever possible, myself and other young instructors would avail ourselves of any opportunities to fly formation. That would then sometimes, whenever possible, lead to "demonstrations" of "air-to-air combat" techniques to our students (not exactly in their training syllabus).

My success in these "tests" of skill against other instructors probably added to my "attitude," along with my frustration at being an "ex" fighter pilot. Don't get this wrong. I think *most* people who knew me then would tell you that I was a nice guy, just somewhat sure of my abilities.

If I thought of a way to demonstrate something better with my actions than my words, I usually did so. I felt that it would be less likely to be forgotten. One example that comes to mind is a demonstration of a "Lazy Eight" maneuver to a student who was having trouble with that particular procedure. I did it with my knees while holding my hands up in plain sight of the student. It was a good one, even though I had never attempted that before. Nothing like putting pressure on yourself. The student therewith decided that if I could do it with my knees, he could surely do it with his hands. He did and had no further problems with transition maneuvers.

Another example I remember is one of which I am not so proud.

It was an instrument training flight with the student in the rear seat of the T-33 jet trainer. He spent most of the flight with the "instrument hood" pulled up over the inside of his canopy to block out any view of the outside world. The only time students were allowed to train on instruments without the hood was when we were actually flying inside of clouds. That particular day was beautiful and clear without a cloud in the sky.

We had been out in the instrument training area and had accomplished all of the basic instrument maneuvers. He had practiced turns, climbs, descents, and slow flight with the landing gear and flaps extended. He had completed several recoveries from unusual attitudes. He had done well. Now it was time to return to the air base and practice some actual instrument approaches.

There was a radio navigational beacon just to the north and west of the airfield. The procedure called for an aircraft to approach the beacon using an instrument inside the cockpit to locate it. Upon passing over the beacon at twenty thousand feet, the student pilot would turn to an outbound magnetic course (printed on the instrument procedure chart from which he was reading) and start a descent away from the beacon and the airfield. During this descent, at a predetermined altitude, he would begin a thirty-degree banked turn back toward the beacon. He would then intercept an inbound magnetic course back to it. He was to level the aircraft at an altitude printed on his chart. This altitude was normally about three thousand feet above the ground.

Upon arriving over the beacon inbound toward the airfield, he would start a timing procedure and a descent to his minimum altitude (a few hundred feet above the ground) and begin looking for the runway. He also had to maintain a programmed magnetic course from the beacon to the field. This was all designed to allow a descent below the clouds to arrive at a point from which he could take over visually (without the instruments) and land.

He arrived over the radio beacon at twenty thousand feet and turned to intercept the outbound course. So far, so good. He began his descent or "Jet Penetration," as it was called. He began his right-hand turn at the prescribed altitude, so far, so good. He leveled off ten thousand feet too high, not so good. (In those days the altimeters were much harder to read than now. They had three moving hands; one indicated tens of feet; one indicated hundreds of feet; and the remaining one indicated thousands of feet. This all of course required some interpretation. Modern altimeters now have a window that displays your thousands of feet numerically.)

I knew what he had done. I let him continue, thinking that he might catch his error. He had leveled at thirteen thousand, five hundred feet instead of three thousand, five hundred. He continued inbound toward the radio beacon. I'm sure that he was feeling good about his approach because he had his magnetic course "nailed." He was also within twenty feet or so of his altitude, he thought. He was actually only within ten thousand and twenty feet or so of his desired altitude.

Remember, he thought that he was only around three thousand feet or so above the ground. I let him go. This was a fairly serious error on his part, and I was hoping that he would catch it. It was not nearly as serious as misreading it in the opposite direction but in actual weather conditions, it could keep him from finding an airfield.

He didn't catch his error. We were over the radio beacon and were cleared for the approach. It was time to begin our descent toward the airfield. I had been trying to decide how to break the news of his misread altimeter to him. But I wanted the information to stick with him for the remainder of his flying career, not just be heard and someday in the future forgotten.

"I've got the aircraft for a second, lieutenant," I said. He "Rogered" and shook the stick slightly, which was the signal to switch control to the other pilot. I shook the stick to acknowledge control and then—remember, this poor guy thought that he was at an altitude of less than three thousand feet above the ground—I rolled the little T-33 upside down and pulled the control stick back in a "Split-S" maneuver (like the last half of a loop). As I did so, I heard the student yell something from the rear seat, and then I heard him almost tearing the instrument hood out of the airplane trying to see.

He thought that he was a dead man. He thought I was a crazy man! The noise stopped suddenly. He had seen how high we were and had checked his altimeter. This time he read it correctly.

I bet he has ever since.

Although the method that I chose to "inform" the student of his error was undoubtedly the most effective I could have used, I was extremely lucky. The poor guy could have pulled the ejection handles and "bailed."

(What's that saying about the end not justifying the means?)

Explosive decompression

Laredo AFB, Texas, 1960

The Air Force treats the subject of Aviation Physiology with great respect. During World War II, aircraft began operating at higher and higher altitudes requiring the use of supplemental oxygen. Even at the relatively low altitude of eight thousand feet, only about half of normal atmospheric pressure is available to assist in the processing of oxygen by our bodies. With the advent of jet aircraft, the requirement for pressurized cockpits and pressurized oxygen supplied to the aircrews became the subject of intensive scientific study and experimentation. The ability of newer aircraft to fly higher and higher had to be matched by systems allowing the crews flying them to survive.

The Air Force requires its aircrews to attend an aviation physiology ground school and altitude chamber training course every three years. A complete review of everything having to do with the hazards of high-altitude flight is accomplished. Any new findings in the field are examined. The perils of smoking, alcohol, lack of exercise, poor diet, and how they compound the effects of high altitude and lack of oxygen are examined.

This course is then culminated by a simulated "flight" in an altitude chamber. Various symptoms of oxygen deprivation are examined. Each participant actually removes the oxygen mask and takes simple written exams while the body becomes "hypoxic," which means suffering from lack of oxygen. The first symptoms are sometimes accompanied by a feeling of euphoria. That phenomenon makes hypoxia even more insidious. A safety observer who is on oxygen watches their progress and helps them to replace their masks when the symptoms are noted. These symptoms can vary between individuals and are very important for the aircrews to record and remember.

At the end of the altitude-chamber "ride," a "rapid decompression" takes place to simulate a sudden loss of pressurization in a high-flying aircraft. Depending on the simulated altitude at the time and the rate of the decompression, the time of useful consciousness is measured in seconds. The pilot must react by donning an oxygen mask and adjusting its regulator to a pressure setting that will ensure his consciousness. He can then rapidly descend the aircraft to a survivable altitude not requiring supplemental oxygen. This is usually around fourteen thousand feet or below.

I was in the T-33 jet trainer alone on one particular night. It was seldom that we got to fly without a student or another instructor with us, and I was really enjoying it. The last time that I had flown by myself was in the F-86F Sabre over a year before.

The moon had just risen over my right shoulder. It had startled me when I first caught a glimpse of it in my peripheral vision. I thought another airplane was about to collide with me. I had snapped my head around so fast that I pinched that little nerve in the side of your neck that feels like a hot poker. I was thinking "Damn, that hurts," and laughing at myself for thinking that I was about to get clobbered by an object around two hundred and forty thousand miles away.

Whoosh! Everything that wasn't fastened down went by my face and out through the soccer-ball sized hole just over my right shoulder. The breath in my lungs rushed out, and the atmosphere inside of the cockpit turned into instant fog. I couldn't even see the lights of the instrument panel.

I had been climbing through thirty-four thousand feet when the hole blew out of the canopy. I was already wearing my oxygen mask, but the oxygen regulator was positioned to a normal setting. I had to get it reset quickly to a pressure setting (forcibly blowing oxygen into my lungs) or I would start becoming hypoxic in a matter of seconds. I fumbled down beside my seat where the oxygen controller was located. I found it and twisted the knob to the full pressure setting. Immediately I could feel the oxygen flowing forcibly into my mask. Some of it was escaping past the edges of my mask, so I tightened it with the straps on its side.

By now the fog had cleared, and I could see the instruments again. It was very cold in the cockpit although only a few seconds had elapsed since the decompression. The minus fifty-seven degree outside temperature didn't take long to invade the formerly cozy warmth of the cockpit. I had to get down quickly and I called the Air Traffic Control Center and told them that I was executing an emergency descent. I switched the IFF (identification friend or foe) to 7700 to indicate an emergency, and down I came with the power in idle and the speed brakes out.

It didn't take long until I was level at ten thousand feet. The cockpit began to warm up slightly, but everything was cold soaked, and the instrument faces became wet with condensation. The weather was clear, and about all I needed was the altimeter and airspeed indicator to land. I wiped them off with my glove and took the T-Bird back to Laredo and landed with no further problems.

I don't know why the canopy failed. (No, it *wasn't* the moon!)

I do know, however, that the training I received in the classroom and the altitude chamber turned a potentially hazardous situation into a near "nonevent."

Thanks again troops!

What next?

Laredo AFB, Texas, 1960

My student was in the front seat. It had gotten dark about the time we had passed Ft. Stockton, Texas. He was flying off the left wing of the lead T-33 in a loose route position about two to three ship widths out and thirty degrees back. I had him flying on lead's left wing because it is a harder position to fly than on the right wing. It has to do with body position. With your right hand on the control stick and your left on the throttle, your body turns naturally to the left. It is a lot more comfortable position than looking to the right and therefore easier to fly formation on the right side. I had found with my first students that if they could learn left-wing well, right-wing was a "piece of cake."

Our weather forecast had been for excellent conditions all the way to our destination, Davis-Monthan AFB at Tucson, Arizona. That's why I was surprised when I began to see lightning flashes in the distant west as we passed out of the glow of lights over El Paso. Our flight leader had seen them too and asked for permission from Air Traffic Control to leave that frequency to check the weather. The flight leader had us change to the Biggs AFB weather frequency with him. We checked in on the new frequency; however, just as the weather forecaster began his update briefing, our radio failed completely. We tried everything including the Guard channel, but to no avail. Our radio was dead. We relayed the visual radio failure signal to our leader with a flashlight, and he acknowledged with his.

We were getting closer to the lightning with every minute. I assumed that the weather briefing must not have alarmed our flight leader because we pressed on westward. As we entered the clouds, the student moved in closer to the wing of our leader. We did not want to lose sight of him and become a "lost wingman," especially without a radio. The air became rougher as we got closer to the lightning flashes ahead. Suddenly there was a bright flash off to the far side of our lead aircraft, and its navigation lights went out.

I grabbed the flight controls and told the student that I had the aircraft. Some additional flashes to our south lit up the lead T-33, and I slid forward and widened out slightly to assure wingtip clearance. Now I could see his orange fuel tank on the tip of his left wing in the glow from the navigation light on my right tip tank. That was *all* I saw for the next several hundred miles. I worked as hard as I've ever worked in my life flying formation. I stayed ready to break away at any time if I needed to, but things were going extraordinarily well.

I had the student keep up with our navigation position as we proceeded westbound in case we ended up on our own. I know it must have been hard for him to look away from the other aircraft's tiptank. With no radio and unknown weather conditions ahead of us, I did not want to be a single ship if I could avoid it. I kept wishing that the lead T-33 would turn around and take us back to El Paso, but I couldn't tell him my wishes. Besides that, I figured that he knew what the weather was ahead and that we'd probably be out of these conditions soon. If that were the case, it was better than doing a one-eighty-degree turn and going back through it.

We were about one hundred miles out of Tucson when the anticipated power reduction came. Lead had signaled with his flashlight, and we started our descent. He was doing a great job, as smooth as silk now that the turbulence had waned. Using flashlight signals again, the landing gear and flaps were lowered. We descended out of the cloud bottoms at about fifteen hundred feet above the ground but it was raining moderately. I matched lead's every move as we flared and landed on his wing.

As we turned off the runway behind him, I could see airplanes parked everywhere. Most of them were B-47s and KC-135 air refueling tankers. I didn't know it at the time but they were there to escape a hurricane which was heading toward Florida.

There was a "Follow-Me" truck with a lighted sign on its rear in front of my leader. (Most Air Force Bases had these vehicles to lead you to your assigned parking area if you were not a base aircraft.) It took off at a rapid clip down an unlighted taxiway. Lead was taxiing ahead of us, and I could just make out his outline in the rain three or four plane lengths in front. We were just passing a B-47 parked off on our right that was pointed toward the taxiway.

Suddenly, the lead T-33's tail lifted and skewed to the left as its nose was jerked to the right. The student and I both stood on the brakes and swerved hard left to miss him. We got stopped without a collision, and we sat there trying to figure out what had just happened.

Finally the Follow-Me truck came back. The instructor pilot and his student had shut down their engine and gotten out of their jet. They were standing in the rain and looking at the right wing of their T-33. Suddenly we were surrounded by flashing lights as the crash and rescue vehicles arrived at our position. In the headlights of one truck, I could see what had happened.

A big B-47 external fuel tank in its wooden crate had been unloaded on the edge of the taxiway and was sitting there unlighted. My leader had rammed his right-hand fuel tip tank right into the side of the crate. That is what jerked him sideways and stopped his aircraft so suddenly. Fortunately there was no fire because his jet's fuel tank and the crated fuel tank were empty; however, there was substantial damage to the lead T-33.

I imagine that there was also "substantial damage" done to the seat of the Follow-Me driver's uniform by his supervisor for first leading us down the unlighted taxiway and then running off and leaving us at a high rate of speed.

I can remember thinking "What next?"

Wanna trade?

Laredo AFB, Texas, 1960

I hadn't thought much about my graduation from Pilot Training at Greenville AFB, Mississippi, before I saw the article on the continuing problems in Air Defense Command's F-104 fleet. The name of one of the participants in the subject accident in the article caused me to think back to that day over a year before. . . .

Aviation Cadets came from many different backgrounds. We had college "dropouts," like me. We had crop dusters. We had truck drivers. We had farmers. We had rich kids, and we had poor kids.

Jim "A." was an exception. He was a highly educated, intelligent aeronautical engineer. He had worked on a project designing and building a vertical takeoff aircraft for Bell before coming to Aviation Cadets. It had huge counter-rotating propellers and was parked on its tail. I'm sure that it was exciting to work on a new project like that, but Jim, like the rest of us, wanted to fly them instead of design them. I could certainly understand that.

Jim was "old." He was at least twenty-five or maybe even twenty-six. But even at that ripe old age he didn't want to fly anything but fighters. I could absolutely identify with that too!

In those days, unlike now, you were allowed to choose your assignment after pilot training depending on your "standing" in the class. All of our flying, academic, and military training scores were totaled and compared with everyone else's. When assignment day arrived, we lined up according to our class standing in a single line. It was formed in front of a desk where sat two or three officers from the personnel section. As your turn came you would step forward, salute, and then be asked to indicate your assignment choice.

There were officers interspersed throughout the line along with my cadet classmates. They were graduates of ROTC (Reserve Officer Training Corps) or commissioned officers who had transferred from their former specialties in the Air Force to Pilot Training. They had been formidable competition for those of us who were Aviation Cadets. They outranked us, of course, and had countless privileges that we did not enjoy. But this was the one place where their officer status did not help them. Only how well you had done in Pilot Training counted there in that line.

I was nervous. There had not been very many assignments to fighters sent down to our particular class. The assignments available to us depended entirely on the "needs of the Air Force." I believe that there were two F-100s (what I really wanted at that time), four F-86Fs, two F-84Fs, and two F-86Ds. Almost everyone wanted to fly "day fighters" as they were called in those days. Those were the F-100s and the F-86Fs. They were basically air-to-air "gun fighters." They also could execute air-to-ground gunnery and bombing as a secondary mission.

The F-84s conversely were primarily used as air-to-ground fighter-bombers. Their lack of performance in air-to-air combat left them at a distinct disadvantage. They were also known to us by the rhyming nickname "Lead sleds."

The F-86Ds we called "Dogs," and they were. Their weapons were guns and some rudimentary missiles. Their engines came equipped with an afterburner for more power, but the jets were so heavy that the afterburner was needed to just get off the ground. The Air Defense Command "owned" these jets and therein was the biggest problem with an assignment to them in my view.

Many new F-86D pilots were being sent right to the top of a mountain somewhere in the world. It was not to gaze at stars and contemplate the universe. It was to gaze into a radarscope at some remote GCI site (Ground Controlled Intercept). There seemed to be a shortage of cockpit seats in the Air Defense Command but not a shortage of GCI-site seats.

I was in line in front of Jim "A." I don't remember my exact class standing but it wasn't number one or two, so the F-100s were out. It appeared that I was going to get the last "day fighter," an F-86F. Jim started "working" on me. He started telling me how much fun it would be to fly F-84s and get to do all that neat air-to-ground stuff. When he saw that it wasn't working, he started extolling the virtues of the F-86D, doggy afterburner and all. I think that he would have written me a check for all the money he had and given me his car to boot if I would agree to not choose the last F-86F.

No way! I saluted and pointed to the last day fighter, an F-86F at Williams AFB, Arizona. The assignment officer dutifully marked a line through it and wrote my name beside it. "*Yes!*" I felt like that five year old in my backyard in Oklahoma.

I stepped aside smiling like an idiot. Jim "A" looked sadly at me before he stepped up to salute and choose his next jet. He picked the first of two F-86Ds at Perrin AFB, Texas. It was over. He was going to fly the F-86 "Dog" and take a chance on not getting an assignment to a GCI site. I was surprised that he didn't take one of the F-84s; I think that I would have chosen one of them if our positions had been reversed.

Graduation day came on October the fourteenth, nineteen fifty-eight. As my mother pinned on my new wings and shiny gold lieutenant's bars, I became a U.S. Air Force Pilot and an officer at the same time. It was quite a series of events for a twenty-year-old. And if that wasn't enough, I was going to get to fly fighters! What a day. I'll never forget it.

We all left Greenville AFB and went our separate ways. While I was busy having a wonderful time at fighter gunnery school, Jim went to the Air Defense Command fighter interceptor school. I didn't give him another thought until. . . .

The F-104 "Starfighter" was brand-new in the Air Force inventory. It was described as "A missile with a man in it."

That little silver sucker would run out to Mach Two (twice the speed of sound) in level flight. It was one of the sexiest looking machines I had ever beheld. I wanted to fly it. (Still do!)

The F-104 had one big engine. It was an awesome jet engine, absolute state of the art. It was both light and powerful. It had the best power-to-engine weight ratio ever. But, the new powerplant was suffering major problems. The F-104s were having engine failures, lots of them.

The manufacturer and the Air Force were "working" on the dilemma, but the engines were still failing at an unacceptable rate.

I picked up an issue of the *Air Force Flying Safety* magazine. Another F-104 had gone down. It was a two-seater version, and both the instructor pilot and the student had ejected successfully. The student's name was James "A." It seems he had graduated from F-86Ds and went straight to F-104s instead of a mountain top somewhere. I thought, "Man, he's lucky. I should have traded!"

You want to know how lucky this guy really was? Within a period of about eighteen months, ex-Cadet James "A." "punched" out of (ejected from) a total of three F-104 Starfighters. Two more and he could have been an Ace! All three bailouts were successful. One of them was in a Starfighter without the newer modified ejection seat. It ejected its pilot downward! That meant if you were low, the jet had better be upside down when you "bailed!"

I wondered after all that excitement if he ever thought about me not trading assignments with him. I know I sure did.

If I had known what would happen to him after F-86Ds and during his early F-104 days, would I have traded him assignments? Of course I would have. I wanted to fly the hot F-104 and after all, *he* made it, didn't he?

Besides that, I was now twenty-two and bulletproof!

If you haven't done it *yet*, you will

Laredo AFB, Texas, 1961

Ever since the advent of aircraft with retractable landing gear, it has been happening: aircraft landing with the gear still retracted.

I can readily understand someone who has flown nothing but fixed-landing-gear aircraft having a real problem. That pilot's mind has been programmed repeatedly that the landing gear is always down and is not a consideration in the preparations for landing. It is not only an excuse for forgetting, but might even be a valid reason.

But not for me. I had never flown an aircraft with anything but retractable landing gear. Every time I had landed, the landing gear had to be lowered. My mind had been programmed that the gear was the most important item in the before-landing checklist; therefore, I didn't have an excuse when *I* almost did "*it.*"

There was an auxiliary landing field east of Laredo. It was used quite extensively. The air traffic was much less dense than the main field, and instructors preferred to take students there to accomplish multiple visual traffic patterns and landings, particularly if a student was having any problems with either. Of course, the takeoff and landing are the most crucial phases of flight, and a student had to be proficient before an instructor could release him to fly solo.

It was a responsibility not to be taken lightly. It was not a pleasant thing to be sitting there watching your student crash and die on his first solo. It had happened to a instructor friend of mine.

This particular student of mine was having some minor problems with traffic patterns and landings in the T-33. It was nothing that I considered serious or "unsolvable," but he needed some work.

We had been practicing pattern after pattern. His landings were okay, but he was having trouble during the turn from downwind to final. He was descending too rapidly in the first part of the turn. Adjusting his descent to maintain a near constant rate until rolling out at the proper altitude on final was giving him "fits"—me too. It was exasperating. (My friend who had to watch his student die had tried to rectify a similar problem with his student. He thought he had the dilemma solved. The solo student lost control and dived into the ground during the base turn from downwind to final.)

That thought was heavy in my mind as I tried everything I knew to get the proper picture in my student's mind. I decided to try one more demonstration with me flying yet another pattern from the rear seat. As we entered the left-hand break (turn to the downwind), I was talking. I was still talking as we rolled out on the downwind. I lowered the flaps and began the base-leg turn, talking. I was talking the student through every move I was making with the stick, rudder, trim, and power. I reduced the throttle still further as the T-Bird began getting a little fast.

The horn started blowing making it hard to be heard so I raised my voice volume on my continuing dissertation on a proper landing pattern. I could hear someone on the radio, but I paid no attention as the damn horn was making it impossible to hear the radio clearly anyway.

As I turned onto final I was still having trouble keeping the airspeed down where it belonged, but it just didn't register why. I pulled the throttle all the way to idle, and we were still too fast. That damn horn was really getting on my nerves.

I saw the red flares shoot out from the runway mobile control tower and realized at about that same instant what the loud horn and the radio had been trying to tell me.

We went back to Laredo.

I checked the landing gear at least three times before we landed.

I have ever since.

I'm sure my student does too.

If you haven't done it *yet*, you will.

The great white hunter

Laredo AFB, Texas, 1961

One of my favorite people at Laredo was a young bachelor instructor pilot named Stan. He was the guy I met the first day on the base, the one calling coyotes in the Bachelor Officers' Quarters.

Stan and I had a few "excellent adventures" together while at Laredo. Whenever we were flying together, something always seemed to happen, and not necessarily for the best.

We liked to take students and fly together during their cross-country training. One Friday afternoon we were going to my former Fighter Gunnery School base at Phoenix, Arizona. We had flown a high-altitude segment of the flight from Laredo and had now descended for a planned low-level navigational route as the last part of the flight. We had dropped down to our low level about forty or fifty miles to the east, northeast of Tucson. We were then headed northwest toward the Coolidge Dam forming the San Carlos Reservoir. We were having a good time charging along at three hundred to five hundred feet above the ground. There was some very spectacular scenery about, and all was well.

We had just departed the area of the Coronado National Forest and descended into a wide valley between some six-or-seven-thousand-foot mountains on either side. Stan had moved out from my left wingtip into a spread route formation. This allowed him and his student to look around without having to concentrate solely on flying close formation. I descended still lower now because I knew Stan could clear himself from the terrain. I was probably at about one hundred feet in the air when suddenly. . . .

Stan made an abrupt pull-up and called "wires" at the same time. It was too late for us to pull over them. Had we been twenty feet higher, we probably would have blacked out power for all of that part of Arizona. We streaked under the wires and probably "streaked" our undershorts at the same time. It was really close. I remember seeing the wires whiz above the T-bird's canopy at around four hundred miles an hour or so.

I immediately made a "command decision" and decided that the low level was over. We climbed back to around ten thousand feet or so and headed for Williams Air Force Base and terra-firma. After we landed, we swore the student pilots to secrecy.

Stan told me that he had been looking across the top of my fuselage at the distant mountain top. He had seen some huge power line towers. He followed the wires with his eyes. When he suddenly realized that they were right in front of us, he had pulled up sharply and tried to warn us.

It's probably fortunate that we didn't have time to try to pull up over them ourselves. I don't think we would have made it and neither did Stan.

They say the good Lord protects fools and children. I don't know which description fit me that day. Probably both!

Sometimes instructor pilots flew together on cross-countries in the same plane. We were required to practice a minimum number of various kinds of instrument approaches while simulating weather flying under the instrument hood or in actual weather. When flying with students, the procedures could not be practiced because the students were not considered qualified as safety observers. We could only fly and "log" approaches with our students if it were actual weather, and we were at the controls.

Once again we had flown out west because both of us liked that part of the United States, and my fiancée lived near Phoenix. We were returning to Laredo on a Sunday afternoon. It was a clear South Texas day. There was hardly any air traffic in that part of the country on Sundays. We had passed Del Rio, Texas, and decided to cancel our IFR clearance (instrument flight rules) and fly the rest of the way VFR (visual flight rules).

I don't remember who was flying in the front seat. It doesn't matter. It could have been me and probably was.

The Rio Grand River naturally forms the border between Texas and Mexico. That border begins in the west near El Paso and ends at the Gulf of Mexico near Brownsville, Texas.

About one hundred miles or so north-northwest of Laredo, a town named Eagle Pass, Texas, is next to the river. That is where we descended into the river bed and stayed there. We flew along merrily between the river banks, which were not very high. We followed the course of the river back and forth as it meandered its way toward Laredo and the Gulf farther on. Hugging the left bank because Mexico was the right bank, we were having a ball.

I don't know why it is so much fun flying fast right down "on the deck," but I've always loved it. I think it may be because it enhances the sensation of speed as you get lower and lower. Also, I think the increased danger causes the adrenalin to pump harder and the feeling of excitement is intensified. I know that not everyone understands what I am describing, but it's real. Race car drivers, sky divers, fighter pilots, we all understand it and crave it. We miss it when it's not there.

We had been down between the river banks for maybe fifty miles when we decided we had better head more directly for Laredo AFB. We came up out of the river bottom but didn't climb much more than that. The sage brush and scrub trees were ripping by in varying shades of green and brown streaks. Suddenly there were some signs of civilization, a dirt road, some fences and an old building of some kind. As we cleared the top of a small rolling hill by a few feet, we roared past some sort of activity so fast that we weren't sure what it was.

It was time to climb. We looked back in the direction we had come but whatever it was had disappeared behind another hill. Laredo was in sight by now and we contacted the control tower and landed.

I was sitting at my table in the squadron area the next morning briefing my students for their upcoming flights of the day. Stan was doing the same at his assigned table. The phone rang, and the operations sergeant called out to Stan and told him to take a call on the nearby phone extension. Stan casually walked over to the phone, lifted the receiver, and answered. Suddenly he clicked his flying boots together and straightened to a position of attention. That got my attention. Stan wasn't like that.

"Yes sir. Yes sir. About fifteen hundred hours, Sir. No sir. Five hundred feet, sir. Yes sir. Yes sir. I'll tell him immediately, sir. Yes sir, I'll pass the word sir! Good-bye Sir!"

Stan's face had gotten redder and redder during his somewhat unilateral telephone conversation. I was really nervous by now because I knew where *I* had been about fifteen hundred hours on the previous day.

Stan hung up the phone and just stood there for a few moments. He then looked up at me and started toward me. He nodded his head for me to follow him. He went outside the flight shack and stopped.

"Do you know who that was?" he asked.

"No, but I bet I can guess his rank from the look on your face. Who was it?" I asked.

"That was *the* man, the Wing Commander," Stan said.

I then said the two short words that almost all pilots say when they think that they're in trouble. (The first word was "*Oh.*")

"What did he say?" I nervously asked.

"Well he asked me if we had been flying yesterday afternoon. He next asked me if we had been to the north of the air base. Then he asked me what time. Following that answer, he asked me how low we were flying and if we remembered seeing any cattle."

"Cattle! What cattle?" I asked.

"The several hundred head in the big corral that we went by," he answered.

"What big corral?" I asked.

"I don't know, I didn't see it either," Stan said. "The colonel then said that he couldn't know for certain if it was us because the caller wasn't sure what kind of jet it was. It went by too low and fast to tell. And, although every one else from Laredo had already landed, he told them that it was possible that it could have been from another base.

"He also told me that he wouldn't debate whether we were at five hundred feet, but that some cowboys on horseback said that they were looking down at the jet when it went by."

"How much trouble are we in?" I hurriedly asked.

"We lucked out. He just gave me some *advice* to pass on to you and the other instructors and students," Stan said.

"It's roundup time on the ranches in South Texas. Those cowhands had just spent several days finding and herding that three hundred head of cattle into the corral. The same corral which the cattle then broke down and stampeded out of when 'that jet' went by.

"Everyone is to keep their altitude above fifteen hundred feet until further notice and to avoid flying over any cattle roundups. Period.

"He also said that it might be a good idea to avoid the small towns anywhere near the large ranches if we happen to be out driving around. And, if we couldn't avoid them for some reason, that we probably shouldn't identify ourselves as jet pilots. We might just get the crap stomped out of us by some pissed-off cowboys."

Stan and I went on another cross-country together that I remember. I had been to our destination Air Force base a couple of years before in an F-86. This time, however, I knew where I was going *before* I got there.

I think maybe Stan had formerly been a student pilot at Webb Air Force Base. While we were there, we went to a gun shop that he knew about, and he bought himself a beautiful little small-bore rifle. He got it with a scope and case, the works. He arranged for the gunshop owner to ship it to him at Laredo. He couldn't wait for it to get there. While he waited, he practiced the wounded rabbit again. I didn't have to hear it this time as I had moved off base into a house with my new wife.

Stan's rifle finally came, and he must have polished it for days. He took it to the rifle range several times to sight it in and practice. It was really accurate. He got to the point where he could consistently knock the "eye" out of the bullseye. In the meantime my wife and I decided to invite Stan over for a home-cooked meal. He jumped at the invitation.

We had a cat named "Chow," a Siamese that was a little bit strange. Stan talked about his new rifle and how he was going to go coyote hunting. He thought that he was almost ready with his "wounded rabbit," his expensive new rifle with scope, and his hours of target practice. He had scouted out some of the areas surrounding Laredo to the north and east and had found the perfect place. It was located on an animal trail. There was a fallen tree where he could sit and wait partially concealed. The view up the trail was clear to where a sharp turn bent around a low scrub tree. It sure sounded like he was ready.

It was time to eat.

After dinner, we went into the little living room and sat down. Chow was very wary of Stan. He didn't necessarily like strangers in "his" house. Stan started to try to make friends with Chow, but the cat wasn't having any part of it.

Stan kept trying to get the cat to come over to him so he could pet him. Chow wasn't interested and kept a vigilant eye on our friend.

Stan started playfully staring at the cat. Either I or my wife told Stan that Chow did not like to be stared at and that he might want to be careful. Stan just laughed and said something to the effect that he had a "way with animals." He kept staring at Chow. I could see that the cat was getting very agitated and mentioned it to Stan again. He still kept staring at Chow.

Suddenly the cat yelled and then Stan yelled when Chow bit him in the leg. Chow had crossed the few feet in a flash and attacked Stan's leg before he could even react.

He wasn't injured badly, but he sure was surprised at the cat's reaction to his "way with animals." Chow had disappeared, and I don't think we saw him again until Stan had gone.

A few days later, after the leg had healed, it was time. It was right at sunset when everything begins to get still and calm. Even the wind seems to choose this time to pause for a while. Stan was at his carefully chosen place beside the animal trail. He hadn't seen or heard anything for quite a while, no activity of any kind. Things were beginning to lose their distinctive edges in the darkening, lengthening shadows. It was *almost* spooky.

Stan kept calling and calling with his best and long-rehearsed wounded rabbit imitation. Nothing. It was getting darker by the minute. The wounded rabbit's plea softly moaned across the sagebrush and through the wind-shaped scrub trees.

BOOM! Like an explosion, the coyote careened around the curve in the animal path. He was at full speed and zeroed in on the probable location of the poor doomed rabbit. He was only a few yards away and closing fast. Stan jumped straight up from his seat on the log and threw his brand-new, polished, and very expensive rifle straight up into the air and then, when he landed, took off in the opposite direction of the coyote.

Stan didn't see the coyote when he crept back to retrieve his weapon, now lying in the dirt. He had made sure to give the coyote plenty of time to leave before he came back unarmed.

I figure that when Stan yelled and jumped straight up, the poor coyote probably had a heart attack and was lying over in the bushes somewhere. Either that, or he had already made it back around the tree in the path from whence he came before Stan even got started good in the opposite direction.

You have got to admire Stan, though. It takes a person with a good self image to have the courage to tell something like that on himself.

You never know. Someone might write a book someday.

The "Great White Hunter" had struck again!

He has a way with animals, you know.

Time kills?

Laredo AFB, Texas, 1961

There are certain items on most aircraft that must be checked for proper operation and position before flight or before landing. If they are overlooked, they may kill you. In fact, I call them the "Killer Items." There are only a few of them. Flight controls, landing gear, flaps, trim tabs, air brakes, power settings, and computed speeds are some that apply to most aircraft.

It sounds like a simple thing, but year after year aircraft and lives are lost because the pilots did not check one or more of these items.

When I was a student pilot at Greenville AFB, Mississippi, there was a young instructor pilot who was a real "go-getter." If I remember correctly, he wanted to be a test pilot. He knew that the more experience that he had, the better chance to achieve his desire. He also knew that the more flying time he had when he eventually applied, the better his chances.

He was impressive. If he was flying with students in the morning, he would fly as many maintenance test hops as possible in the afternoon. If he was instructing students in the afternoon, he would be out at sunrise taking off on another maintenance test flight. He knew what he wanted and was willing to work for it.

There was a young instructor pilot at Laredo who was much like the "go-getter" at Greenville. He would fly test flight after test flight. They did not have to call anyone else. He would fly as many in a row as they had ready to go. He had a procedure that he used to save time. He would preflight all of the airplanes that were ready to be flown at the same time. Then he would come back from each flight and jump right into the next already preflighted jet and go test fly it. I remember thinking more than once that I hoped that he wasn't getting in too much of a hurry to build his flying time and experience.

He had already flown one or two test flights and was preparing for the next one. As he was waiting for clearance to taxi onto the runway for takeoff, he rapidly went through his checklist. Everything obviously appeared correct because he taxied onto the runway when cleared and then began his takeoff when authorized for that.

I don't know when he realized that something was amiss. I doubt if he ever knew exactly what was wrong with his T-33 jet trainer. He was solo in the little jet when it began to roll to the left. He wasn't very high, only a few hundred feet. He wasn't very far from the end of the runway, where he had just made his last takeoff.

As the jet continued its slow roll to the left, I'm sure that he took the normal corrective action and moved the control stick to the right.

The T-33 continued in its roll to the left unaffected by his control-stick inputs. He undoubtedly pushed the stick all the way to the right, to no avail. He was such a good pilot that I imagine he even tried moving the stick in the opposite direction just in case the aileron controls had been inadvertently reversed—they weren't.

The T-Bird was almost in ninety degrees of bank when he gave up his fight to control it and pulled the ejection handles. Everything in the ejection sequence worked as designed trying to save the eager young pilot's life.

He had waited too long. He hadn't given the ejection system a chance. He was still in his seat when they got to him. The ejection had shot him out nearly horizontally with the ground because of the jet's steep bank angle. The seat vector was too low.

He had been doomed when he pulled the ejection seat handles. He had been doomed when he did not notice that the ailerons were rigged wrong. No, they were not reversed. Instead, when the control stick was pushed in one direction, they both went up. When it was pushed in the other, they both went down.

I hope pilots reading this will immediately ask why he didn't use the rudder. Just as you do, I think it could have allowed him to control the airplane. In fact, I know it. Perhaps he could not have landed, but he would have been able to control the jet high enough to safely bail out.

Yes, "Time" killed him. First, he didn't *use* enough. Then, he didn't *have* enough.

Don't rush through those "Killer Items."

Fear of flying

Laredo AFB, Texas, 1961

Toward the end of my second year at Laredo, we began to receive the T-37 jet trainers. The Air Force was closing the Primary Flying Schools and consolidating all the pilot training at the Basic Flight Schools, including Laredo. Basic was operated and manned by Air Force personnel. There would no longer be any civilian instructors in the Air Force Pilot Training program. I think it had to do with the advent of jets in Primary training and the expense of sound-proofing the flightline buildings for the sirenlike whine of the T-37 engines.

The noise was unbelievable. There was a simple fact that was driven into us. If you didn't wear ear protectors when they were running, you would lose your hearing in those ranges of sound. It was unnecessary to tell us that, though. The sound was so painful that you couldn't bear it without some sort of protection. Can you believe that the Air Force would buy an airplane that caused all those problems? There's no telling what had to be spent at each host base to soundproof buildings to try to protect people.

I still have trouble understanding why they didn't just tell the manufacturer that there wouldn't be any sales until the damned airplane was quieter. You don't suppose that politicians were involved, do you? Nah, they wouldn't do that, would they? I've got about two hundred and forty hours in the T-37, and I still detest the thing. And they're still flying around to this day, driving people crazy with their noise.

Noise wasn't the only problem the Air Force was having with the T-37. It seems that they were losing quite a few of them in accidents. One of the requirements in the pilot training curriculum was spin recovery techniques. Students were not allowed to spin the aircraft alone, but instructors were to demonstrate spins and the recovery from same. Unfortunately, sometimes the recovery wouldn't work, with predictable results. The manufacturer had supposedly fixed the problem with a three- or four-inch-wide strip of metal attached around the nose parallel to the fuselage. That did not fix the problem, and they were still spinning into the ground.

After two years instructing in T-33s, I was selected for the Check Section. Remember the hatchet men I talked about at Hondo? I had become one. Basically all I did was give check flights to student pilots. Unfortunately, I gave them in both the T-33 and the T-37. The Air Force decided that a few instructors from each base would be selected to go to a T-37 "Spin School" that had been established at Craig AFB, Alabama. After a few weeks of spinning and recovering, they would return and instruct the T-37 instructors in the new spin recovery techniques that had been instigated. The Air Force had ceased spin recovery instruction until an instructor had received the new training by the "spin instructors." Guess who got picked?

For the next three months, all I seemed to be doing was spinning the T-37: upright, accelerated, inverted, inverted accelerated. Climb to twenty thousand feet and spin down to eight or ten thousand. Climb back up and do it again and again and again. I especially "enjoyed" the inverted spins. Hanging upside down in your harness with the earth spinning around seemingly above your head was great "fun," especially with no pressurization, another T-37 "plus." It was particularly enjoyable when the guy you were instructing got it wrong and you had to take over the controls with barely time to make the recovery. It wasn't easy, and it wasn't fun! I wager that two hundred of my T-37 hours were spent spinning.

One evening I was eating dinner. It was some sort of a chicken dish, and I remember thinking that I must have swallowed a bone because my lower throat began hurting every time that I would swallow. Later, after I went to bed and laid on my back, I felt something hot on my chest. I turned on the light, and there was nothing there. I laid back down, and the sensation returned. I rolled onto my side, and it disappeared. After several weeks of this pain and burning sensation, I decided to go to the Flight Surgeon.

The doctor listened to my story. After I had finished, he looked over my medical records and then gave me a short physical exam including an x-ray. He told me to return several days later for the results, which I did. He explained that he had found nothing wrong and would send me to Lackland Air Force Base for more tests the following week. About a week after I returned from Lackland, he called me and asked me to come in. In the meantime, my pain had suddenly disappeared as quickly as it had come. I sat down in his office, and he began to tell me that all the tests had shown nothing physically wrong. I was confused and concerned and told him that I hadn't imagined the pain.

He then said that he thought he knew what was wrong with me. "Lieutenant Cook, your pain was psychological. You have a fear of spinning the T-37."

I saw nothing, but red! I couldn't remember being so angry. This little pompous ass had the audacity to tell a fighter pilot that he had a fear of flying! I stood up suddenly and bent over his desk. I guess I must have looked menacing because he leaned back so quickly that he almost tipped his chair over backward. No, I didn't hit him. I just grabbed my medical records from his desk and stalked out the door. I went straight down the hall to the office of another flight surgeon. I marched into his office, and he saw the look on my face and asked me what in the world was wrong. I told him my story, and he took my records from me and began to look over the tests and what the other doctor had written.

After a few minutes he closed my medical records and looked up. "Jerry, I can tell you what was wrong with you. You say you have been doing spin instruction in the T-37?"

I nodded, and he added, "Have you been pulling a lot of negative Gs?" I nodded yes again. He then said, "There's a diaphragm between your stomach and your esophagus. It keeps stomach acid out. Your esophagus is not equipped to handle acid and burns when exposed to it. All of the repeated negative Gs you've been pulling over the past months finally tore a hole in that diaphragm. When you ate, the swallowing action caused some acid to squirt into your throat. When you laid down on your back, a small amount seeped into your lower throat. While all these tests were being made and studied, your body healed itself. You're good as new."

I looked at him and said, "If you can tell that from what he wrote, why couldn't he?"

He said, "I'd rather not answer that." That gave me all the answer I needed.

Flight surgeons, like pilots in those days, were required to fly four hours a month in order to collect their flight pay. Flight surgeons were not pilots. They would ride in the right seat of a T-37 or the back seat of a T-33 in order to obtain their four hours. One of my good friends was the favorite pilot of the doctor who had diagnosed my "fear of flying," more specifically, my fear of spinning the T-37. I went to my pilot friend and asked him if I might fly the flight surgeon. He said sure, he didn't like to fly him anyway.

I then told him that I wanted it to be a surprise and that I wanted to borrow one of his name tags for my flight suit. He thought that it sounded like a good joke and readily agreed to let me fly the flight incognito. The day arrived, and I preflighted the T-37 early and got in to await my passenger. He showed up in a few minutes and merrily waved at me and began strapping in the right seat. I had my tinted visor pulled down with my oxygen mask on and my friend's nametag on my flightsuit.

The flight surgeon plugged in his helmet connections as I started the engines and said "Hi, Charley."

I just grunted a reply and accomplished the checklist. As we taxied for takeoff, he continued his "small talk," and I continued my short replies. We took off and began climbing to twenty thousand feet in the T-37 transition area. He was just looking around and enjoying the ride. When we got to our altitude, I made a couple of clearing turns and pulled the throttles to idle and the nose up and held it there.

He looked at me and said, "Charley what are you doing? You know I don't like stalls." I didn't answer. When the plane stalled, I pulled the stick full back and pushed full rudder. The little bird began spinning. I shoved full opposite rudder and jammed the stick full forward. We were now spinning upside down. I then released some forward pressure from the stick and we began our inverted, accelerated spin. We were whirling around upside down at about a revolution per second.

By now the doctor had a death grip on the top of the instrument panel shroud and was screaming at me at the top of his lungs, "Get it out, get it out!" I waited as long as I dared and finally began the involved recovery procedure. We pulled out of our dive at about three thousand feet. He jerked off his mask and began barfing in one of his gloves which he had removed.

When he had finally finished, he turned and looked at me with wild eyes. "Charley are you crazy?" he said.

I raised my dark visor and unhooked my oxygen mask. The incredulous look on his face as he recognized me was totally rewarding.

I said, "Do you still think I'm afraid to spin the T-37?"

His eyes suddenly blazed, "Lieutenant, I'll have you court marshalled for this!"

"For what?" I replied. "All I did was an authorized procedure that I do every day. I just thought a flight surgeon like yourself would want to experience what the pilots go through. Isn't that part of your job, to know what pressures and stresses we experience so you can better diagnose our problems? Besides that, I'm sure the board would be interested to hear that another doctor was able to take your tests and written comments and correctly diagnose my case when you couldn't, *doctor.* By the way, do *you* have a fear of spinning the T-37?" He didn't say another word.

Every time I saw him after that on the base he would avoid me like the plague.

By the way, the next time I saw Charley, he said, "What in the hell did you do to the doc? He said he didn't want to fly with me anymore. Then, when I told him it was just a joke, he said it wasn't funny."

When I told Charley the whole story, he thought it was hilarious and said that he didn't like to fly the "little bastard" anyway.

Failure, fun and higher learning

Laredo AFB, Texas, 1961

I only encountered one student that I couldn't help as an instructor or as a check pilot. Even though I was assigned to Check Section to conduct student check rides, I sometimes would decide to call it an instructional ride if I saw a problem that I thought I could remedy. I was to give a progress check to a student whose instructor had given up on him. His grade book indicated that he was way behind his peers and had not yet soloed. The check ride was to evaluate him to see if I thought there was any potential for him to continue.

I had empathy for students who flew with Check Section because of what had happened to me at Hondo. I always did everything I could to get the student pilots to relax and feel at ease. It usually worked, but this student was really "clanked." I felt sorry for him, but I could only do so much. We briefed the flight and proceeded to the

T-37. He was as white as a sheet, and his hands were shaking. Nothing that I said or did seemed to help.

Finally, it was time to take off. We were airborne, and he climbed to about fifty feet above the ground and leveled off. He had the throttles at maximum power and the landing gear and flaps were still down. I waited as long as I could before suggesting that he raise them. I was concerned that we would damage them if we went any faster with them extended. He raised the gear and flaps but did not reduce the throttle settings or climb. It was summer, and the air was hot and bumpy. We went faster and faster and bounced harder and harder.

After a couple of miles like that, I suggested that we make our ninety-degree turn to the left to stay in the traffic pattern. He immediately stood the little T-37 on its left wingtip and pulled about four Gs. He rolled out on an easterly heading, and we still hadn't climbed. We whizzed by the VOR station southeast of the airbase. The airplane was going as fast as it would go. Finally, I looked over at the student. His eyes were wide and sweat was rolling down his face as he fought to control the plane.

I said "Lieutenant, is anything wrong?"

He finally blurted out "Sir, I'm lost!"

"Maybe if you would climb up to where we belong, you could see where you are." I knew that we were only a few miles southeast of Laredo AFB.

He buried the stick in his stomach, and we must have pulled six Gs. The plane shot straight up, and he froze. Finally, when it ran out of speed, I took the controls and recovered from our unusual attitude. As much as I hated it, I recommended that this young man be eliminated from pilot training. He was so frightened of flying that it must have been a nightmare for him, but he didn't want to quit. You had to admire him for his courage, but he just wasn't meant to fly as a pilot. He was transferred to navigator training, and I understand that he did quite well in that endeavor.

Another student provided even more excitement and a definite learning experience for both of us. He was having trouble landing and had not soloed because of it. His problem seemed to be judging his distance above the runway. He simply would not get down to the proper height to begin his flare for landing. His instructor had even sent him to the flight surgeon to have his depth perception rechecked. The student was otherwise very sharp. He flew us to the auxiliary airfield with precision and expertise. He hardly lost or gained any altitude in any of his traffic patterns. His airspeed control and flight control techniques were outstanding. He just wouldn't get down to the runway low enough to land.

We had been going around and around the traffic pattern for what seemed like forever. We only had enough fuel for one more approach and landing, before we would have to return to Laredo for our final landing. I was exasperated. His flying otherwise was outstanding, and I couldn't discern his problem. We were coming down final for about the twentieth time. Everything was looking good. He descended right through the altitude where he had been leveling off before, and I thought "He's finally got it." We were almost to the point where we should begin our landing flare, and he suddenly pulled the nose up and leveled off again at his prior altitude.

I lost my cool! I smacked the stick forward with my hand. He left it there! The nose of the T-37 pointed dangerously at the runway. I yelled "Don't give up!" as I grabbed the controls to salvage the landing. *He raised the landing gear handle!* Just as we were about to touch down, *he raised the landing gear!* I rammed the throttles full forward and pulled the nose up. We were barely above the runway and just above a stall. The runway was about seven thousand feet long and we skimmed along about a foot or two above it, trying to regain flying speed.

I thought about slapping the gear handle back down, but I was afraid the gear doors would hit or the gear would drag on the runway before it could fully extend. The end of the runway was fast approaching, along with the fence beyond it. I very cautiously eased the stick back, and we cleared the fence with inches to spare. Finally I got the T-37 flying again, and we started to climb away from the airfield. The student was looking straight ahead with tears in his eyes.

I cleared my throat and said, "Lieutenant, why did you do that?"

He turned to me and blurted out, "Sir, I thought you said gear up!" Bless his heart, he was trying so hard to please me that he mistook my "give up" for "gear up."

I then said, "What would you have done if you thought I had said 'blow up'?"

"Sir, I guess I would have tried," he answered.

I turned the airplane back over to him, and we started back to Laredo AFB. He performed a perfect traffic pattern and a perfect landing. I followed a hunch and told his instructor that I thought he was ready for solo. He did just that on his next flight. He graduated very high in his class and went on to fly F-102s in the Air Defense Command.

I still don't know why he couldn't get down to the runway before, but evidently our little episode "fixed" the problem.

Me? I decided to return to Mr. Kurkendal's calm, quiet instructional techniques and watch very carefully my choice of words, especially when flying with students who were eager to please.

The screaming mimi

Laredo AFB, Texas, 1961

I thoroughly disliked the airplane. It was ugly. It was slow. It was designed with side-by-side seating like a bomber or an airliner or something. I had never sat side by side in an aircraft before except on the way to pilot training in the back of an old DC-3.

And then there was the noise! It was almost unbearable! It even filtered in from the outside when you were flying and sounded like a siren, but that's in another story.

One day I was in the T-37 transition area with a student. We were completing our aerobatic training for the day with a roll or two. He seemed to be having some kind of trouble entering and exiting the aileron rolls. He complained about the ailerons, and I took the aircraft to check it out. It felt normal to me in a couple of turns. The student said it felt funny when he had to put in more aileron to roll. It also felt normal to him during the clearing turns he made prior to the rolls.

I lowered the nose and accelerated to the recommended airspeed for aileron rolls. I then raised the nose and rolled rapidly toward the left. (Fighter pilots almost always roll left if there's a choice. I don't know exactly why. It just feels better.) I tried to move the stick back toward the right to stop the roll as we approached wings level upright. It wouldn't move. We just kept *going* and *going* and *going*, kind of like that pink rabbit. Every time, as we approached upright, I would pull back on the stick to keep from losing excessive altitude.

I was straining to pull the stick back to the center, but it wouldn't budge. This was getting real old by now. I told the student to get on the controls with me, and we both pulled as hard as we could. It felt like something was going to break. It did.

The rolling stopped. The student let go of the controls, and we looked at each other. The aileron mode felt sloppy. I looked out at the right wing. (I was in the right seat.) The right aileron was bouncing up and down a little, like it wasn't attached—it wasn't.

I flew the little "screamer" back to Laredo and did it with great care. I didn't use over five or ten degrees of bank at the most and flew a long straight final for our traffic pattern.

I don't remember exactly what had occurred, but part of the aileron mechanism in the area between the wing and leading edge of the aileron had jammed. We had finally broken the mechanism and pulled the jam loose.

In the manufacturer's defense, we were *extremely* hard on their equipment. They build fine aircraft, and I have flown several of their designs and still do.

It also speaks very well for them that many T-37s are still flying around in the United States Air Force more than thirty years later.

Fine feathered friends?

Laredo AFB, Texas, 1962

Jet aircraft are a marvel of engineering. They have great integrity of strength built into them when they are stressed in the normal methods and directions. Both the Lockheed T-33 and Cessna's little T-37 were designed to take many negative and positive G forces and had safety factors built into them exceeding the limits to which we adhered.

Two of my instructor friends were killed by an outside force that stressed their aircraft beyond the design limits. Following are two different accidents, one in each of the aforementioned aircraft, each by a warm-blooded but dangerous adversary, a bird.

In the first accident, two instructors were flying together. It was a local-area checkout for the newly assigned pilot in the back cockpit of the T-33 Shooting Star. They had begun the orientation flight an hour or so before and were just minutes from returning to Laredo for landing.

Their altitude was about three thousand feet above the ground. They were several miles to the south of the air base, just a few miles east of the Rio Grande. They had been viewing the student training areas with the newer pilot in the back checking his local area map from time to time. They were both looking down at some distinguishing feature on the ground for future navigational reference. The front-seat pilot looked up and through the forward windscreen of the jet trainer.

A big black object was rushing right at his face. He instinctively ducked as the airflow rushing up and over the contour of the jet's nose lifted the object just clear of the canopy. The airflow then pulled the buzzard down, and it skimmed the fuselage until its body met with the bottom of the vertical stabilizer where it rises from the aft aircraft body. The forward area of the stabilizer is very thin at that point. One would think that the big bird would just have been cut in two and swept on past the vertical fin.

The force of the bird's body against the bottom area of the stabilizer was later estimated to be in the tens of thousands of pounds per square inch. The impact sheared the vertical stabilizer right off of the T-33. It immediately began to slew sideways as it lost directional control. The aircraft then entered into what was described by the front-seat pilot as a flat spin. There was no controlling the aircraft.

The frontseat pilot gave the bail out command and pulled the ejection handles. He evidently was followed a few seconds later by the backseater.

According to the report I heard, the frontseat pilot's parachute opened and almost immediately he was on the ground. The backseater's parachute was out of its pack and deployed, just microseconds from blossoming and arresting its occupant's fall.

The flight surgeon and accident investigators said just a half second or so more and he would have made it. They could tell by how the chute streamed out behind his body. He had impacted the ground in a near kneeling position with his head and his knees striking the ground at the same moment. Just one instant more and his body would have been snapped upright and slowed by the blossoming parachute. He probably would have walked away. Instead, his broken and lifeless body was flown away by the rescue helicopter.

The remains of the bird on the separated vertical stabilizer were analyzed and found to be that of a buzzard. It was a plentiful resident of that area. They estimated his weight, and with that, and the speed of the jet, they were able to calculate the tremendous force exerted by the scavenger. One of the experts said it was like a knife through hot butter.

I would have never thought it possible, but it's true.

For the second bird-strike fatal accident, it was a hot, lazy summer afternoon. The Texas sun was merciless as it penetrated the atmosphere and heated everything it touched beyond bare-skin tolerance.

It was slightly cooler where the little T-37 jet rolled and looped at twelve thousand feet. But it was still about eighty degrees, still hot and sticky in the dual cockpit.

The student had only recently made his first solo flight, and his instructor was instilling in him some of the basics of aircraft aerobatics.

They had begun with several clearing turns, which are the ninety-degree turns in two directions so that they could visually scan the surrounding airspace for other aircraft. An old aviation axiom states, "Look around, a midair collision could spoil your whole day."

They had accomplished aileron rolls, barrel rolls, loops, Cuban eights, Immelmanns, lazy eights, spilt Ss, the works. Between each aerobatic maneuver, they had carefully checked for other aircraft.

The fuel gauges in the trainer indicated that it was time to head back to Laredo for a couple of overhead traffic patterns and landings and then a cold drink in a cool room. As they descended toward the airfield from the transition practice area, they again conscientiously performed their clearing turns. A constant concern is letting down on top of another aircraft directly below you in your blind spot.

Their descent complete, they were only a few miles from Laredo and landing. They were heading toward the common traffic pattern entry point, which is a visual point over which most aircraft that are headed for a particular runway, pass at a specified altitude. The entry point is like a "gateway" through which to enter the pattern.

The vulture slammed into the windshield on the right side of center. Barely slowed by the impact, it struck my friend full in the face and snapped his head back. His helmet visor was shattered. His oxygen mask was ripped from his face. His neck was broken.

The student at first did not realize what had happened. It had seemed like something had exploded in the cockpit. He had been flying with his helmet visor lowered, therefore his eyes were protected. He was covered with bits of flesh and bone. Blood was spattered everywhere.

His instructor had not responded to his calls over the interphone. The wind noise was horrendous. Its force through the shattered windscreen had blown everything not secured all over the cockpit. Maps and checklists had been either rammed behind or beneath the ejection seats. Some loose items had been ejected from the jet by the howling wind's swirling action.

The student's initial confusion began to wane somewhat. Slowly he began to realize what had happened and that he was still alive and functional. He was the only one.

The inexperienced fledgling pilot was flying "solo" again. He didn't realize that his instructor, only twelve inches or so to his right, was dead. He just knew that the motionless pilot was seriously injured, and he had to get them both on the ground.

Although it was difficult because of the wind noise and his own shock, the student pilot was able to communicate with the runway mobile control tower. He landed the aircraft.

The fire trucks and ambulances had responded at his call, and they formed a circle around the now still little jet on the runway. As some of them assisted the shocked student pilot in exiting his aircraft, others gently lifted the instructor's lifeless body from his ejection seat. With some of them, their tears flowed freely. Others struggled hard to choke them back. But there was no "manly" shame for having shed tears. There was only human grief and a deep shocked sadness at what they saw.

A midair collision had done much more than spoil a day.

It had "spoiled" a promising young life.

True class

Laredo AFB, Texas, 1962

I was really enjoying being a Standardization-Evaluation check pilot. All I had to do was fly with instructors. Student training missions were behind me.

Each rated instructor pilot at Laredo had to pass two flight check rides a year. One was to test their instrument proficiency, the other to check their instructor techniques and general flying prowess. Flying these flights with those pilots was the best flying job at Laredo. They were all excellent pilots and instructed students on a daily basis in the same areas that were checked by me. It was easy.

I was a first lieutenant. It's not exactly at the bottom of the rank "totem pole," but close. I flew with pilots of various ranks all the way up to lieutenant colonel. Because of my job as their evaluator, I was responsible for the flight and the individual "accountable for the aircraft." It was a pleasant position for a young lieutenant to be in. All of the higher ranking people that I checked managed the situation very well. I respected their rank, and they respected my position and responsibilities.

I guess that my supervisor felt that it was time for me to gain a little more experience for my flight checking résumé. He scheduled me to fly a check ride with none other than our Wing Commander who was a full colonel. I had only met him once in a receiving line at his New Year's Day reception, and I had been so "clanked" that I forgot my wife's name when I tried to introduce her. (I really did!)

I can remember how nervous I was that evening. You would have thought that I was the one getting a check ride. I suppose one could say that I probably was, in a way.

Colonel "Smith" was very personable during the briefing, and I know in retrospect that he was trying to put me at ease. I made it through the preflight briefing and began to relax a little in response to the colonel's manner. It was his annual instrument check ride to determine if he was competent to fly on instruments in weather whenever the need arose. All Air Force pilots have to prove each and every year their continued instrument flight qualification. (Different "Smith" characters might appear from time to time out of respect for all affected parties.—Ed.)

The colonel made a simulated weather takeoff (ITO) under the instrument hood, and we headed to San Antonio to accomplish some instrument approaches before returning to Laredo for our final approach and landing. We always took the pilots to other places than their home base because of their overfamiliarity with the Laredo instrument approach procedures. That would have been too easy for them and would not have been a good test of their skills.

We arrived in the San Antonio area and were under the direction of San Antonio Approach Control. Our first procedure was to be a VOR (visual omnirange) jet penetration and VOR low approach to Kelly AFB. Then we would make a GCA (radar controlled approach) and an ILS (instrument landing system) approach before returning to Laredo.

There were two VOR stations just a few miles apart at San Antonio. The one to which we had navigated on our flight to the area was the San Antonio VOR. The one that we were supposed to use to execute the jet penetration was the Kelly VOR. If you utilized the wrong one, you would be several miles from your proper position and would not arrive where you are supposed to be. In mountainous terrain or if there is a tower in the way, it's a killer. All that has to be wrong is four little numbers improperly entered in a navigational radio.

The colonel did not catch his error, and I was not supposed to assist him except if safety was involved and it wasn't, yet. I kept hoping that he would see his mistake before we arrived over the VOR and began our penetration from twenty thousand feet to around four thousand. If he didn't, I couldn't let him continue and would have to intervene. I couldn't believe that I was in such a situation. I felt like I might be "damned if I did and damned if I didn't."

Actually the decision was an easy one to make. I had no choice but to do my job. As he turned outbound for the penetration, it was too late for him to catch his error. I told him that I had the aircraft and then made a turn toward the Kelly VOR station. I informed him of what he had done and waited for his reaction.

"Well lieutenant, I sure screwed that up. What do you want to do?" he asked.

"Sir, let's go on and complete the mission as briefed, and we'll talk about it when we get back to Laredo, if that's all right with you, of course," I said.

"You're the boss tonight. Let's do it," he replied.

All during the remainder of his check ride I wondered what I was going to say after we landed. He did not make any more errors and flew an excellent flight. That seemed to make my job even harder. I knew what I would have to do for anyone else. I would either fail them on the flight if they made any more serious errors, or just have them repeat the failed procedure on a later flight if they did not.

We walked into the briefing room together and sat down. I still was searching for the right words.

"Lieutenant, I know that you probably feel that you're between a 'rock and a hard place.' I know that I made a serious error back there. You waited as long as you could giving me a chance to catch my mistake. I did not. Fair is fair. Can you schedule me for my recheck tomorrow night?" he asked.

It was like a huge load was lifted from my shoulders. I quickly explained to him that a complete check ride was not necessary, just another jet penetration at another location and that was it. I debriefed him on the rest of his flight.

We finished the rest of his check ride the following night. It was excellent. Colonel "Smith" was a class act.

He was an officer *and* a gentleman.

Instructor follies

Laredo AFB, Texas, 1962–63

Beer "call." Academic instructors were poor unfortunates. They were instructor pilots who had been selected, or "screwed" as they put it, to teach the student pilots their ground courses. In that day's Air Force, pilots as well as flight surgeons were required to fly for a minimum of four hours per calendar month to qualify for their flight pay. Toward the end of each month, there was usually a lot of scrambling to get the academic guys on the flight schedule.

That particular night, there were some student solo cross-country flights scheduled. One T-33 jet trainer aircraft was assigned to orbit each of two checkpoints at the far corners of a triangular four-hundred-and-fifty-mile course. The job of the crew in the orbiting aircraft was to ensure every student airplane passed each checkpoint and to log the time of passage. It was the last night of the month, and an academic instructor was assigned to the back seat of one of the checkpoint aircraft.

When starting the engine, a frontseater called the bad news to the backseat occupant: "We have to abort. We have no EGT indication. The exhaust gas temperature gauge is a must item for flight." Not wanting to lose his flight pay, the academic instructor immediately called the other checkpoint plane and its backseater readily agreed to give up his seat on the boring flight. The backseat pilot switch was rapidly accomplished because the engine was already running on his new bird. Without any further delay, the T-33, already slightly late, took off and hurried to its orbit point over Alice, Texas.

They arrived just as the first student pilot checked by with his position, time, and altitude. The orbit aircraft answered, "Roger, Checkpoint One is twelve thousand, five hundred feet, left orbit, north of Alice."

Suddenly, an alarmed call, "Checkpoint One, this is Checkpoint Two, say again your position!"

"Roger, I'm at twelve thousand, five hundred feet, just north, northwest of Alice."

"Roger One, this is Two, I'm at the same approximate position, heads up, I don't see you!"

Frantically, the Checkpoint One frontseat pilot looked around the sky, checking in every direction. As he did so, the backseat pilot in Checkpoint Two did the same. They were both about to panic, fearing a midair collision.

Finally, the Checkpoint Two backseater called over the interphone to the frontseater, "Do you see Checkpoint One anywhere?"

The answer "I am Checkpoint One" immediately came back to him!

As the realization of what had happened came over them, their laughter began. They instantly made a pact not to tell anyone, but it was too late. Three student pilots had figured it out before they did, and the word beat them back to Laredo.

The beer that the academic instructor had to buy for everyone eagerly awaiting his return cost most of that month's flight pay. The next month, he flew early!

Along for the ride. After about three years at Laredo, I was assigned to be a Standardization-Evaluation pilot. I was very pleased with the assignment, as it was considered the elite job in the Air Training Command. We only flew with instructor pilots. We administered to them their two annual checkrides.

The South Texas skies were blue with high fleecy clouds at about twenty thousand feet. The instructor pilot in the rear seat of the T-33 jet trainer was attempting to demonstrate one of the more difficult coordination maneuvers, called a Lazy Eight. When it is performed properly, it is a beautiful maneuver with constantly changing bank, pitch, and airspeed. To fly one perfectly is nearly impossible. The backseater was accomplishing his annual proficiency flight check. Because his primary duty was as a flight instructor, his checkride also tested his instructional techniques in addition to his flying ability.

Jerry and I were good friends, but I could tell that he was nervous. It was a common occurrence with almost everyone when they flew with Standardization-Evaluation check pilots, which was my primary duty. He was botching up his Lazy Eight demonstration, and after his third attempt, I told him to relax a little and I tried a couple for kicks. I thought maybe it would help him just to see someone else attempt them. After the second one, I said "Just do whatever you want while we reduce our fuel down to landing weight." He said "Okay." I released the stick, laid my arms along the canopy rails, and began humming some mindless tune to myself.

The Texas landscape grew larger, then smaller as the nose of the jet slowly fell, then rose again as it increased airspeed. This was accompanied by the wings increasing their angle of bank with each gentle maneuver. As I looked around the practice area, clearing the sky around us for other airplanes, I noticed that the changes between nose up and nose down were getting much more pronounced. After one particularly nose-high, low-airspeed situation, followed by a very nose-low, high-airspeed dive toward Texas, I casually said, "Jerry, what kind of maneuver are you doing?"

His immediate, startled response came back, "ME?"

I learned a valuable lesson I've never forgotten. Make sure *someone* is flying the airplane and both pilots know who!

What good eyes. Another of my duties as a member of the Standardization-Evaluation section was to give recurrency flights to pilots transitioning back into the T-33 jet trainer from other types of aircraft. My "student" for the day was an F-100 fighter pilot just transferred back to the Air Training Command. Our takeoff was scheduled for about 1 P.M. on a sunny summer day. I was to ride the back seat. After a few transition and aerobatic maneuvers designed to redevelop his feel for the T-33, we would then practice numerous landings of all types.

The preflight and engine start went normally, as did the initial part of the takeoff; however, just after landing gear retraction, I felt a slight airframe vibration. Suspecting that I knew the most likely reason for the problem, I asked to take control of the aircraft. After gaining adequate maneuvering speed and a safe altitude, I rolled the plane into a eighty-degree right bank and looked down.

"Yeah, just what I thought," I said. "We have a landing gear door hanging out just a little." I recycled the landing gear and rolled the aircraft eighty degrees again. Looking down I said, "No, it's still hanging out a couple of inches, but not enough to damage anything if we keep our speed down." I could see the pilot in the front seat looking down also and shaking his head slowly. I gave control of the T-33 back to him and we finished our flight, modifying it slightly to hold down our maximum speed and G forces.

After the flight and after debriefing all aspects of the mission, I asked the other pilot if he had any questions. "Well, yes," he said. "When you were banked steeply and looking down, could you really see the gear door hanging out?"

"Yes," I said.

"Damn, you must have good eyes, all I could see was a little shadow of the airplane on the ground. He continued, "I couldn't see any details at all from that altitude."

Suddenly I understood. "I wasn't looking at the ground three thousand feet down," I said. "I was only looking about fifteen feet away, at the shadow of the airplane's belly on the tiptank!"

"I knew I shouldn't have asked," he said. "I knew it! And I bet you won't tell a soul will you?" he continued.

"Nah, I wouldn't do that, now would I?" I answered.

Fickle finger. Chuck and I were going cross-country. Since Laredo was one hundred and fifty miles from anywhere, it was a real treat to get to fly somewhere interesting and get away from students for a couple of days. We had both already flown a couple of student training flights that day and were looking forward to our weekend away from South Texas. I had probably twelve hundred hours in the T-33 by now and had come to know quite a lot about it, including a few tricks. Some of them I could use to check whether students had accomplished an item on the checklist. For example, I could tell if a student had turned on his landing lights or only the single taxi light by looking at the amperage load. There were several other useful ways to ascertain if the instruments were being scanned properly.

I decided to try one of them on Chuck, not because I didn't trust him, but because I was already bored riding in the rear seat on this portion of the flight. We had only just taken off and were climbing out still in sight of Laredo. I placed my right index finger against the glass front of the exhaust gas temperature indicator on the rear panel. I had learned that if you rubbed up and down rapidly on the face of the instrument that it would cause static electricity. The temperature-indicating needle would begin bouncing up and down. The static electricity would travel through the wiring to the indicator on the front panel and cause it to bounce up and down also, although to a lesser degree. Since we were flying a single-engine jet, that could be a quite disturbing indication.

I was having a ball rubbing the exhaust gas temperature gauge and, over Chuck's right shoulder, watching the front one bounce. The amber engine overheat light illuminated! I just stared at it for a couple of seconds and then at my finger, with which I had quit rubbing. Then I thought, "I know!"

If one pulled down very carefully on the fire warning test switch in the front cockpit, it could cause either the amber overheat or the red fire light to illuminate without the other. They both came on together if you pulled the switch down normally. Not everyone knew about that particular trick, but I figured that Chuck did and was just getting even with me. I stretched and looked. He was not touching it. The engine overheat light was real. We landed and spent the weekend in Laredo.

I still wonder about that!

Laredo two-step. The radio came to life suddenly. I hadn't expected anyone to call in for another twenty to thirty minutes. The morning launch of T-33s had ended only five minutes ago, and I was catching up on the log books for which I, the mobile control officer for the day, was responsible.

"Carter, this is Skidrow." Carter was the callsign of the center runway mobile tower.

"Go ahead Skidrow, Carter," I answered.

"Roger, request a straight-in full stop landing." That was an unusual request, especially since Skidrow had only been airborne for fifteen minutes according to the logbook.

"Skidrow, you are cleared straight-in full stop. Are you having any problems," I asked.

"Nothing serious, just a small physiological one," he answered.

Thinking the student had become ill, I asked, "Do you want me to have the ambulance standing by?"

After a long silence, "Not unless it has a 'crapper' in it," came the somewhat terse reply.

Skidrow landed and turned off only halfway down the seven thousand foot runway, no small feat for a T-33 still heavy with fuel. As I watched with the field glasses, the T-bird slowed in front of Base Operations, the canopy was raised, and the instructor, not the student, came scurrying over the rear canopy rail, down the back of the wing, and off on the ramp. As I watched, he half ran, half shuffled into the building. All the while, he was pressing the back of his flightsuit tight against his backside.

When I saw him later and asked him if he made it, Skidrow said "Almost" and grinned sheepishly. (By the way Skidrow, how's everything in Baton Rouge?)

Laredo was definitely a "learning experience" for me. In retrospect, I realize that my days of instructing were more valuable than I could ever appreciate at the time. Watching others make errors and then having to demonstrate the correct and precise way to fly probably made me a better pilot than I would ever have been otherwise.

There were many more experiences that I could relate from Laredo, but I want to move on, back to my first love, fighters.

Besides, instructing students was too dangerous. At least you know the enemy is going to try to kill you.

Still a mystery

Laredo AFB, Texas, 1963

Instructor pilots were not getting assignments from the Air Training Command to fighters. The closest you could come was the RF-101 Voodoo reconnaissance fighter, and there were very few of them available. I had just about resigned myself to the probability that I was never going to get to fly fighters again. Instructors were mainly getting assignments to bombers, airlift, and air refueling tankers.

I came back into the Stan-Eval section one day and began hanging up my parachute and helmet. My supervisor walked up to me with a strange look on his face and asked, "Who do you know?"

I said "What are you talking about?" He handed me a piece of paper and I began to read it. It was an assignment for me, and it was to MacDill AFB, Florida. I was going to be flying the F-4C, presently the world's hottest fighter! I couldn't believe my eyes. I sat down and reread it again and again.

"Jerry, you must know somebody. This doesn't just happen," my supervisor said.

"I swear major, I don't know anyone and haven't talked to anyone about assignments," I replied.

"Well if you figure out how this happened, please let me know so I can get one too," he continued. I was absolutely in shock. You couldn't have gotten the grin off my face with a bulldozer. F-4 Phantom IIs! Zero to thirty nine thousand feet in one minute and seventeen seconds. One thousand, six hundred miles an hour. Two and one-half times the speed of sound. I knew all the numbers. I had been reading everything about them that I could find, and now I was going to be flying them! They were so new to the Air Force that they were just now going into production. I was assigned to the first wing in the Air Force to get them and I do not know, to this day, how it happened. No one on the base believed that I didn't know how I got the assignment.

You should have heard some of the offers I got to trade assignments. I was offered everything from cars to their firstborn. But I wouldn't have traded that assignment for anything. I was hanging on to that paper for dear life. In fact, I was afraid that it had been a mistake and that any day I would be getting another piece of paper stating that fact. I remained concerned about that possibility until the day I checked in at MacDill.

One a day in Tampa Bay

MacDill AFB, Florida, 1963

From "Laredo By the Sea" I found myself at Tampa by the bay. It was amazing to behold the contrast! It was from sagebrush to palm trees. It was from sleepy little South Texas town, to thriving Florida city. Instead of brown rolling hills, there were blue Gulf waters. Instead of fifteen-year-old, four-hundred-mile-an-hour training jets, it was sixteen-hundred-mile-an-hour screaming fighters that hadn't even come off the assembly line yet. It was a dream come true and one that I hadn't even dared to dream. Where that awesome assignment came from I have never found out, and I was still fearful from time to time that I would wake up and find that I was still at Laredo Air Force Base spinning the T-37.

MacDill had been a Strategic Air Command base before it became a Tactical Air Command fighter base. B-47s had once been stationed there. They were a notoriously underpowered aircraft when they were heavily loaded. Evidently several of them had not been able to remain airborne after taking off over the bay at the southwest end of the runway—and so the saying "One a day in Tampa bay." It was unbelievable. They sometimes had trouble just staying airborne after takeoff with their power-to-weight ratio. On the other hand, an F-4 after takeoff could be at an altitude of twelve thousand feet by the end of the runway. It made me think of some of my classmates from F-86 school who were still flying B-47s. That made my situation seem even more phenomenal and me even more grateful. Whoever you were who gave me that assignment, thank you again!

There were no Air Force F-4Cs at MacDill or elsewhere in the Air Force yet. The first one was still being finished at the plant in St. Louis; however, there were several Navy F-4Bs parked on one of our ramps. They were on loan from the Navy for the checkout school. The instructors were Air Force and were commanded by Colonel "Speedy Pete" Everest of test pilot fame. He was well known and held at least one speed record from his Edwards Air Force Base days. I was really eager to get into one of those beasts and do some speeding myself, but first, academics again.

As I had learned about seven or eight years earlier at Oklahoma University, academics about a topic in which you are interested are a breeze compared to the stagnant air of an uninteresting subject. Electrical and hydraulic systems were suddenly fascinating. New terms such as pneudraulics were eagerly put to memory. F-4 performance figures and capabilities were totally absorbed. I felt like I couldn't wait to put them to the test.

November 1963 was an eventful time for me. About the middle of the month I got my first F-4 flight and on the twenty-sixth my first child arrived. Jeffrey Wayne Cook was born at the MacDill Air Force Base hospital. Jeff probably didn't have a chance of not being a pilot. I imagine the first sound he ever heard was an F-4 in afterburner. His baby book has a section about things that were happening at the time of his birth, such as who was the president, etc. One of them was "The fastest airplane in the world." Of course it says "My daddy's F-4 Phantom II!"

Other pilots think that fighter pilots are crazy. They say that we take off in the first place with minimum fuel, then set fire to it. They're probably right, but setting fire to an F-4 with the afterburners and feeling that surge is worth being crazy. My first flight in an F-4 lasted almost the same amount of time as my first flight as a pilot in the little T-34 just six years earlier. Exactly seventeen minutes after takeoff we were touching down. We had lifted off and crossed over the end of the runway in a near vertical climb at twelve thousand feet. We had leveled off at thirty-five thousand feet in less than two minutes. I had then joined the "Mach 2 Club" out over the Gulf of Mexico west of St. Petersburg, Florida. Now, seventeen minutes after takeoff and some two hundred and fifty miles later, we were landing back at MacDill. Welcome to the world of high-performance fighters. I loved it. I felt like I had finally arrived.

School days

MacDill AFB, Florida, 1963

It was the most enjoyable but also the most intense and demanding training that I had received. The F-4 was a tremendous advance in technology over the other fighters of the day. It was capable of multiple roles like no other plane had ever been before. It could perform air defense, air-to-air combat, interdiction, conventional bombing, and nuclear weapons deliveries with equal aplomb. But it could only do these things if we learned how to use it properly. It was a far cry from a T-37 or T-33, and it was almost twice as fast as the Air

Force's present frontline fighter, the F-100 Super Sabre. In a word, it was awesome!

The control sticks in the rear seats of the F-4s we had borrowed from the Navy were removable. Sometime during our training we were required to fly in the backseat and direct an air-to-air intercept. That was to help us understand more about the rearseat pilot's job, which also consisted of being a radar intercept officer. I remember the sinking feeling when the crew chief came up the ladder and took my stick away. Pilots don't like flying when they don't have a set of controls; however, the radar scope in the Navy F-4 extended out into the space occupied by the control stick in the rear; therefore, no rearseat stick during air intercept missions. The Air Force F-4s did not require the stick to be removed to use the radar, but they were still in St. Louis being assembled.

The intercept training mission out over the Gulf of Mexico went smoothly, and we had "shot down" the target aircraft several times. We had reached our bingo fuel state, which meant that it was time to go home. As we turned back toward the shore of West Florida, I stowed the radar scope to get it out of the way and to give me more room. The instructor in the front seat said, "Jerry, do you want to fly us back?"

I looked down at the floor between my boots and could see the slot fitting where the stick was normally attached. It stuck up a couple of inches and was moving around as the pilot manipulated the controls. I bent over and touched it and realized that I could still see over his left shoulder. Obviously, he had forgotten that I didn't have a stick. Just for the hell of it, and to see if I could do it, I said "Sure, I've got it."

He repeated "You've got it" and wiggled his control stick. He then put his arms up on the front canopy rails and away we went toward MacDill. It was hard to move the stick socket, but I could do it. I had to bend over just enough to reach it and still see out. I moved the throttles as needed and the rudder pedals to help me when I rolled in and out of turns. We entered the traffic pattern and sped toward the end of the runway at fifteen hundred feet above the ground and three hundred knots. I knew that I wouldn't have enough mechanical advantage to accomplish the break for landing, and I had no trim.

I said, "You'd better take it now."

The frontseat pilot said "Why? You can land it from back there if you want to."

"Not without a stick," I replied. He jumped like he was shot as it registered and he remembered the mission we were on.

"How the hell?" he blurted out as he grabbed the controls and yelled "I got it." We porpoised a couple of times before he settled down and entered our fighter break over the end of the runway.

After we parked and shut down the engines, the frontseater hurriedly unstrapped and came back to where I was sitting filling out the aircraft forms. He looked over the side of the cockpit at the socket on the floor and repeated, "How the hell?" He then shook his head and exclaimed, "Well shit, I'm glad you didn't want to land it!"

Another mission I flew in school comes to mind. It was a great ego booster and confidence builder for me. The flight was a two-ship of F-4s assigned to accomplish what was called "one v one" air combat tactics. My instructor, who for that flight was also my opponent, was a colonel, and combat veteran. By his order, following that mission some thirty-two years ago, he shall remain anonymous.

After a wing takeoff and formation flight out to the Gulf and our practice area, we turned forty-five degrees away from each other and separated ourselves by about fifty miles. At his call, we were to turn toward each other and locate the "enemy" with our radars. After closing to visual range, we were then to engage in what we called a "furball" or "dogfight" using visual procedures. When he called for our turn toward each other, I rolled inverted and dove straight for the Gulf waters thirty-five thousand feet below.

As we pulled out of our dive at around five hundred feet and six hundred knots, my backseat pilot located the colonel's F-4 on our radar about thirty-five miles south of us and some seven miles above. I maintained my speed just below Mach One. At what I estimated was the right point, I lit the afterburners and began an expanded Immelmann maneuver, basically the first half of a loop with a half roll to level flight at the top.

As we reached the vertical position, I saw the other F-4's belly as it passed over us. It was almost perfect. I rolled out at one-half mile in trail with him and my gunsight was on his tailpipes. If it had been for real, he would be going down in flames. As I closed the distance between us, he was banking back and forth searching for us. I joined up two aircraft lengths behind him and matched his every maneuver. Finally his voice came over the radio. "Two, where are you? Are you lost?"

"Not unless you are sir," just slipped out before I could stop it. I knew I had just "let my mouth overload my ass," so to speak. I finally said "I'm right behind you sir," in my most apologetic tone. He banked hard left, and I saw him look back at me.

"All right two, join up and let's go home. Lead is bingo." I looked down at my fuel gauges and knew that he wasn't low on fuel. I figured that he was probably getting low on something else, like his "temper control." All the way back to MacDill, the colonel never looked over at me the way a flight leader usually does to check on his wingman. Every time I looked at his backseater, he gave me "the finger."

I dreaded the debriefing. We all stood up when the colonel walked in the room. He looked at me and said, "Captain Cook, I guess you're feeling pretty smart about now."

"Actually sir, I feel pretty stupid," I replied.

"That too," he said. Then he said, "I don't know how you did what you did, but it was obviously effective. I'm going to leave now and you can brief my backseater who, by the way, got his butt chewed out because he couldn't find you on our radar. He can brief me later. But one more thing. You did *not* beat me in a dogfight, and I don't expect to hear anything otherwise outside this room. Understand?" Only one answer was obvious.

"Yes sir," I said. "Perfectly." I think he was grinning when he left the briefing room, but I'm not sure. We never flew in the same flight again.

After I debriefed the backseater on what we had done to "disappear" from his radar, he just gave me "the finger" again.

A lifesaver, a killer

MacDill AFB, Florida, 1964

The ejection seat in the F-4 was amazingly effective. It was also amazingly complex. It had to be capable of lifting its occupant clear of the aircraft at speeds from one hundred knots (later rocket-assisted versions were capable of zero airspeed ejections) up to extremely high indicated airspeeds. It had to be "smart" enough to separate the pilot from his protective seat immediately and cause the parachute to deploy instantly at low altitudes. It had to "know," when to delay the separation of pilot and seat at higher altitudes to protect him from the hostile elements of cold, speed, and oxygen deprivation. It would then begin seat separation and parachute deployment at the proper altitude unless overridden by the pilot.

The Phantom II's amazing variance in speed and altitude capabilities dictated the complicated design of its ejection seat. That intricacy came with a price.

A pilot from our squadron was walking by the hangar when he heard the detonation. He knew it couldn't be good news as he ran toward the open hangar door along with several maintenance personnel. Inside the hangar were several F-4s in various states of disassembly.

People were gathering in two different parts of the building near a Phantom with a canopy missing. The news spread quickly. It was not good.

Two egress technicians had been performing maintenance on one of the ejection seats in their assigned F-4. One of the technicians was giving the newer airman some tips on the seat. He was demonstrating how to service the "banana links" as they were called. The area of interest was located on top of the seat and very critical to proper operation. It was also a very hazardous section because any movement of the links could initiate the seat-ejection sequence.

The version of the tragedy I heard was that the senior technician was cautioning of the dangers involved in servicing the area when the ejection seat fired. He was struck in the upper body and catapulted to the top of the hangar where the seat hit before falling back to the concrete. The other maintenance technician was knocked to the hangar floor and seriously injured.

The critical "banana links" had evidently been inadvertently moved. The result was one fatality and one seriously injured.

In another instance, a Phantom was cleared onto the runway for takeoff. The rearseat pilot read the checklist item "Canopies closed and locked." As the front canopy started down, a broken part of the mechanism contacted the critical area on top of the seat, and the poor unfortunate unsuspecting aircraft commander was ejected into his own canopy and killed.

A new procedure was begun immediately after this terrible accident. The rearseat pilot was to check the condition of the front canopy as it was lowered and warn the front seater to avoid the lowering sequence if everything was not normal.

To my knowledge that particular problem never occurred again.

The ejection seat in the F-4 saved the lives of many of my friends and acquaintances. It was a remarkable and effective addition to the extraordinary capabilities of the Phantom II. But like the great machine of which it was a part, it was deserving of great respect and prudence.

But then, this is a dangerous "business."

Receiver ready!

MacDill AFB, Florida, 1964

Man we were "hot." You couldn't touch us with your bare hand or you'd probably get burned. We were jockeying the hottest fighter in the world. It was twice as fast as the Air Force's former frontline fighter, the F-100 Super Sabre. We were still very new in the jet and one of the hardest things was locating the proper switches in the busy cockpit. We were working hard to fulfill all the myriad requirements to become combat ready. One of the requirements was air-to-air refueling, both day and night.

As you will observe in a subsequent story, the Phantom II could be extremely sensitive to large rapid control inputs. It was my first daytime air refueling sortie. I was number four in a four-ship formation of brand-new shiny F-4s fresh out of the factory in St. Louis, Missouri. We had met our tanker somewhere over the Everglades in south Florida. The joinup turn had been accomplished, and we were headed back northbound on the air refueling track. The track was an imaginary oval in the sky, over one hundred miles long and several miles wide. It was administered by the Air Force and controlled by the air route traffic control system at Miami. Thus, air traffic separation between us and civilian aircraft was provided by the radar controllers. Separation among ourselves was our own responsibility.

The procedure for "formating" with the KC-135 Strato-Tanker was for the F-4 formation to join up on the tanker's right wing. The fighters would be in an echelon to the right, which meant that the F-4 leader would be immediately off the right wingtip on the KC-135 with numbers two, three, and four flying off the right wingtips of lead, two, and three. No one, of course, was on the right wing of number four as he was the last aircraft. After refueling, the fighter aircraft would then form on the left wing of the KC-135 in a left-hand echelon.

Back to the story. One at a time, beginning with the formation leader, we would move from our right-wing location to a position approximately fifty feet behind and below the KC-135. This was called the "precontact position." After moving to this position, the frontseater had to reach down near the throttle quadrant for the switch that opened his air refueling door. This also served to depressurize the fuel system, which allowed the aircraft to accept fuel. It also turned on a small green "ready" light beside the pilot's windscreen.

Our flight leader smoothly moved into the precontact position while the rest of us remained on the KC-135's right wing and watched intently. The tanker boom operator—he lies on his stomach in the rear of the KC-135 and controls the air refueling boom—called "Tanker ready." This meant that he was ready for the fighter to move forward toward the air tanker and then stabilize in the proper position for the refueling. This movement was guided by a system of two rows of "director lights" on the belly of the KC-135. They were controlled by the boom operator until the aircraft were connected together by the refueling boom. Their directions to the fighter pilot then became automatic, driven by the fighter's position on the air-refueling boom.

The boom operator then called, "Sir, your air-refueling door is still closed." I saw the lead pilot suddenly duck his head looking for the air-refueling switch in the maze of other switches in the still unfamiliar cockpit. The nose of his F-4 dropped simultaneously with his head dropping. His head suddenly popped upright, and the nose of the Phantom popped up also. He pushed the stick forward to keep away from the tanker, and the nose dropped again. The pilot jerked the stick back again and the F-4 stalled and began a "snap roll" to the right. Fuel corkscrewed out of the fuel dump valves on the trailing edges of the wings and from the fuel vent mast below the rudder. Sometime during this wild display of prowess, the air refueling door opened. (The pilot had found and moved the switch just as the "air-show" began.) He quickly recovered the Phantom and powered immediately back to his former position. He then said, "Receiver ready." Is that cool or what?

The boom operator had never seen an F-4 before. So of course this was also his first experience with refueling them. He jokingly asked, "Do you have to do that to get the refueling door open?"

The super-cool fighter pilot replied, "Only occasionally."

There will be no accidents!

MacDill AFB, Florida, 1964

I had missed the briefing, so the information I had was secondhand. But according to my sources, the auditorium was filled with nearly two wings of new F-4 pilots. On the stage was a rocking chair. A new general in the Tactical Air Command was about to give a briefing.

The Tactical Air Command commander had been a fighter pilot's fighter pilot. Everyone liked and respected him. Most of his senior staff were like him, and they had reflected his philosophy. This new general in the command was an "unknown" to the Phantom pilots assembled in the room. He was formerly a bomber pilot from the Strategic Air Command.

There was a lot of good- and not-so-good-natured rivalry between the two combat commands. There weren't very many fighter pilots who thought a bomber pilot could fly for "sour apples." Conversely, most bomber pilots thought fighter jocks were all egotistical, immature flyboys. (We always wondered what was wrong with that?)

The gathering was called to attention, and the new multistarred general appeared on stage and sat down. After everyone was seated, the briefing was started. One of the subjects was the Air Force's decision to crew its F-4s with two pilots instead of a pilot and a radar intercept officer. The Navy had been using the one pilot and RIO combination since they began operating the Phantom II several years prior. They believed that a specially trained radar intercept officer who was not a pilot was the way to go. They felt that particular combination was most effective for their mission.

According to the general, the Air Force's approach was different. They felt that having two pilots was the best method with the backseater being cross-trained in RIO duties. In case of an incapacitated frontseater, the backseat pilot could take over and return and land the aircraft. Sounds good doesn't it?

Only a couple of "minor" problems here. The backseater could not lower the landing gear by the normal system (no landing gear handle). The backseater could not stop the aircraft because he had no brakes. He could not select afterburner if he needed it and could not shut down the engines. (Rear seat throttles only moved between idle and full power without afterburners.) This could prove rather tricky when certain hydraulically powered systems were lost when lowering the landing gear by the emergency system with the front seat pilot's gear handle still in the "up" position. (Frontseater incapacitated, remember?) In this condition, the lower engine bay auxiliary air doors stayed open. The engines would then suck hot air in through the doors with the aircraft on the ground, and the engines would automatically reset their idle speeds. They could reset them to near full military power, and the backseater could not stop the jet. (Bear in mind, no brakes.) These are the items that I can remember after all these years. There may have been more. Add to these the lack of visibility from the rear seat, and you've probably got an accident or an incident waiting to happen.

I guess the Air Force figured that was better than the probable fate of the crew if the backseater was *not* a pilot. I'm not debating whether the Air Force was correct or not in its decision for crewing the F-4. (Later, however, the USAF started putting radar observer/navigators in the backseat. You can reach your own conclusions.)

My problem was with the general's own purported statement about all this. Some of those present told me he said something along these lines, "With two big engines and two pilots, there will be no excuses for F-4 accidents. Anyone who has one will answer to me personally."

I'm not saying that I or any other pilot is superstitious, but most of us figured a statement like that was the "kiss of death," so to speak. Sure enough, it was only a few weeks until one of our sister squadrons "lost" an F-4 on the Avon Park gunnery range. The pilots didn't have to answer to the general, though. They were dead. Their Phantom had "departed from controlled flight" for some unknown reason at a very low altitude and impacted the ground.

Some weeks later, a very good friend of mine who lived just down the street was flying cross-country in a Phantom II. He was on instruments in clouds when a loud explosion occurred. It was somewhere deep within his jet. The control stick moved full aft of its own volition and, of course, the nose pitched up. The F-4 went out of control. He tried to take corrective action, but the control stick was immovable. He stayed with the aircraft longer than he should have trying to recover. Finally, he gave the bailout order and both pilots ejected successfully.

The word trickled down that the accident investigation board found the pilots not at fault, but cited some unknown massive mechanical failure. That sounded pretty feasible to me. How about you?

Suddenly the aircraft commander was removed from his status as an F-4 fighter pilot and spent the rest of his career flying the old, slow T-33s. As far as I know, he never got to fly fighters again. It was so grossly unfair and such a supreme waste of talent. Once again it didn't pay to be first.

We didn't realize it at the time but we were to hear more from the "no-excuse" general while we were in Vietnam.

Ours is not to question why! (Don't bother politicians with facts.)

MacDill AFB, Florida, 1964

Ah, the "Whiz Kids!" The Secretary of Defense's answer to a military person's prayers. They were some of Harvard's and Yale's former "finest." Now they were in control of the Pentagon, and they knew everything 'bout 'most everything.

The guy looked sharp in his business suit and expensive shoes. He reportedly was one of the "Top Guns" from the Defense Department. He and several "clones" had arrived in an Air Force jet and were met by our local brass. Obviously, he was somebody. He also looked to be about my age, twenty-seven, maybe younger.

One of my squadron's fighter pilots had his orders for the day, wear a new flight suit with polished boots and stand static display in one of the hangars with one of our new F-4 Phantom IIs. It was parked there without a speck of dirt or oil anywhere, surrounded by a vast array of the weapons that it was capable of carrying. It was indeed impressive. My fellow fighter pilot was there to answer any questions that the visitor might have. He didn't have any. He knew it all. He was impressing his entourage of carbon copies with his vast knowledge. He informed everyone, including the fighter pilot, that he was a pilot himself and owned his own airplane. He also assured those in trail with him that he could pilot one of these F-4s, if he had been so inclined.

As he circled the aircraft with its weapons, he patted his hand on one of the three-hundred-and-seventy-gallon fuel drop tanks suspended under each wing. As he did so, he turned to his followers and said with much authority, "I have been told that these Sparrow missiles aren't worth a damn." The Sparrows were on trailers several feet away and behind the expert. Enough said. He then went back to Washington to report to whomever. Probably another Whiz Kid at least twenty-nine or thirty years old.

A few weeks later, our commander called a meeting to brief us on our "hush-hush" mission. It seems that the President was soon to be traveling to West Palm Beach. The Cubans had been making threats, and extra precautions were being taken. There would be two Air Force Ones. No one would know which one was our Commander in Chief's. The fear was that Cuba would fly an aircraft to Florida and try to eliminate the President. Our job? To protect him with our new F-4s.

Only one problem. We had no weapons. We had no missiles. We had no guns. All we had, that I was aware of, were the inert static displays. So new were our airplanes, that we had no training with actual weapons as yet.

Word was that our local commanders tried to point out to Washington the facts about our lack of combat readiness. It was my understanding, that the "White House" insisted on the President being protected by the "fastest fighters available," and that was us. It didn't seem to matter that there was an F-100 wing at Homestead AFB, south of Miami. It made no difference that they were combat ready with all the required weapons available. After all, they were *only* about twice as fast as the best that Cuba had at the time. We were three times as fast and brand-new, so we must be better. Right? *Right!*

The day came. We flew the mission. The President landed unharmed. The Cubans didn't show.

I'll leave it to your imagination as to what our tactics were to be if they had.

Armed Forces Day

MacDill AFB, Florida, 1964

You'll recall that I had been assigned to the first wing in the Air Force to receive the F-4. We had spent the first few months of 1964 riding to St. Louis on TWA and picking up new F-4s at the McDonnell factory. We were getting to fly more and more as we received our quota of airplanes and were building our proficiency and weapons qualifications. It was early spring and requests were rolling in from all across the country from Air Force bases wanting us to bring F-4s to them, both for static display and flight demonstrations on Armed Forces Day.

One of the requests was from Hill AFB, Utah. It was the home of the huge supply depot that housed and stored all the parts for F-4s. None of those personnel had ever seen the airplane that they were supporting, and their local commander thought that they should. Our commander agreed, and straws were drawn to select a crew. One of our flight commanders, an older crusty major, was selected as the aircraft commander. He seemed like he didn't want to go and griped about it at every opportunity. He downplayed the whole thing and said he thought it was unnecessary to go all the way to Utah to show them an F-4.

I'm sure that we weren't supposed to hear any of what happened when they arrived at Hill AFB, but we did. Their arrival had been timed to coincide with the end of the work shift for the day at the depot. After a very-low-altitude high-speed pass followed by a vertical climb, they entered the traffic pattern and made a slow-speed low approach and then executed a tight closed pattern and landing. Remember, there wasn't another aircraft around that could perform like an F-4.

As they came taxiing in, they could see a huge crowd of several hundred people there to watch them. The wing commander was there to greet them in his staff car. They began to think maybe this was a big deal for these folks. The frontseat pilot taxied smartly up in front of base operations and stopped abruptly with the nose strut bouncing impressively. As he shut down the engines and began to unstrap himself, the backseater exited his seat, walked back atop the left engine intake and down onto the wing. He then stepped down on the fuel droptank and onto the ground. Everyone was watching every move with great interest.

The aircraft commander had finished unstrapping and stood halfway up in the cockpit before he realized that there was no ladder to climb down like at MacDill. He was going to have to use the built-in steps and integral sliding ladder in the side of the fuselage. He suddenly remembered that he had never used them before and had never paid any attention as to their location. Close to a thousand people were now watching his every move. After all, he was the highly skilled fighter pilot of this Phantom II, that they had just seen perform with such deftness and expertise. They were intrigued.

The backseat pilot on the ground had also become very interested in his fellow pilot's actions. He watched with thinly veiled amusement as the frontseater threw his foot over the canopy rail and guided his shiny boot down the black line leading to the first kickstep. Success. The first foot was in. Now for the second. When he finally located the next black line, he realized that he had the wrong foot in the top step. If he continued, he would be cross-legged on the side of the F-4, and that wouldn't work at all. All of the audience was watching intently now, including the backseat pilot, who by now was enjoying himself immensely. The aircraft commander had no choice. He hoisted himself by his arms and quickly reversed his legs. As he placed his second foot into the second kickstep, it worked as advertised and the hidden two-step ladder slid out of the aircraft side and clanked into place. Finally, something was going right.

The pilot took his top foot out of its kickstep and stretched down toward the top ladder step before he realized it. He was stuck. The "fickle middle finger" of one glove had caught on something on the canopy rail and would not come loose. As he tried to pull free, the middle finger of the glove just got stretched out longer and longer. The pilot on the ground looked at the curious crowd and began to clap. The onlookers joined in and the applause was thunderous. Finally, the stranded pilot climbed back up, freed his glove, and slowly descended from the Phantom in the proper prescribed manner. When he reached the ground, he first glared at the other pilot, then took a bow as the crowd went wild.

Guess who told us.

My Armed Forces Day at McCoy AFB, Florida, wasn't nearly as entertaining as the one just described, but it did have its moments. I was standing at ease in front of my F-4. The sun was hot, and the heat waves were coming up from the ramp. Next to me was an easel with a large poster containing all the statistics about my Phantom. No one ever read it as far as I could tell. They would walk up to me, look at the airplane, and then start asking me questions. Fortunately, I knew most of the answers. One kid asked me what I thought was a strange question, until I looked back over my shoulder where he was looking. His question was, "Will that tailhook come down if the safety pin is in it?" As I started to answer him, I turned around to see another kid swinging on the tailhook. It wasn't supposed to come down, but with a three-thousand pound pressure charge on it, I didn't want to let him test it. It would have driven him right into the ramp if it had actuated. I politely got the kid away from the hook and became somewhat more vigilant for the rest of the day.

Right after I shooed away the kid, a man walked up to me. He was wearing a three-piece suit and had a cigar in his mouth. I looked carefully at the cigar and was relieved that it wasn't lit. He looked at my F-4 and said "What kind of aircraft is this?"

I said "It's an F-4C, sir." He then looked at the fighter parked next to mine.

"What kind of airplane is that?" he asked.

"It's an F-100, sir," I replied.

"What does your airplane do?" he asked.

"It's a fighter," I answered.

"What does that one do?" he asked.

"It's a fighter, too," I said.

"How fast is your plane?" he continued.

"One thousand, six hundred miles an hour, sir," I said.

"How fast is that one?" he said.

"It will fly eight hundred and fifty miles an hour," I replied.

"You mean yours is twice as fast as that one?" he asked, astonished.

"Yes sir, it is," I said proudly.

"Then what the hell did we buy that one for?" he said. He stalked off shaking his head, not the least bit impressed.

Do you think he was a taxpayer?

One other incident comes to mind from that day at McCoy AFB. I was really surprised to see the two older ladies walking toward me. One of them looked to be around seventy and the older one must have been ninety or better. I figured them for mother and daughter. Here they were, out in the hot sun at this air base looking at airplanes. I was amazed. They came walking right up to me, and the older one stopped in front of me and looked me over from head to toe. She was all bent over and only came up to about my chest. She had a cane in one hand. She then looked past me at the F-4 sitting there.

"What kind of airplane is this sonny?" she asked.

"It's an Air Force fighter called an F-4 Phantom II, ma'am," I replied.

"How fast does it go?" she continued.

"It has a top speed of just over sixteen hundred miles an hour," I answered.

"One thousand, six hundred miles an hour! My goodness! Are you the pilot?" she asked.

"Yes ma'am," I said as she looked me up and down again.

"Well, what does your mother think about you going that fast young man?" she scolded.

I tried not to smile as I informed her that my mother didn't like it a whole lot.

"I should think not," she said, then shuffled away down the hot ramp.

Mom's max performance climb

MacDill AFB, Florida, 1964

I knew my mother didn't like my flying an F-4 Phantom as fast as one thousand six hundred miles an hour. I was so sure not because of something my mother ever said, but because of what she never said.

My parents had driven down to our home in Tampa to see our new son Jeffrey Wayne. They were excited because he was their first grandchild, and they now had someone to carry on the family name. As I was an only child, there had been a good chance that it could end with me.

I was scheduled to fly the day after they arrived, and I had asked my wife to drive them out by the end of the runway at MacDill Air Force Base just before I took off. I knew that I would be able to see them and could wave to them to let them know which one I was. What was even better, in my eyes, was the fact that I was taking off as a single aircraft and not in a formation. That meant I could probably get the tower to let me execute a maximum performance takeoff in my Phantom II, which was quite an impressive thing to watch. At that time, the F-4 Phantom II held all of the world's speed and climb records.

The last opportunity they had to watch me take off was several years earlier in the old F-86 Sabre. It really wasn't that impressive with just over five thousand pounds of engine thrust. The F-4 had thirty-four thousand pounds of engine thrust, and it showed. I was really looking forward to impressing my parents. I figured that watching me takeoff almost straight up would do just that. I felt like I had when I used to try to impress them by going fast on my new bike as a kid. (Actually I guess it wasn't that much different. Just a little faster bike.)

I stopped my jet in the number one position immediately short of the runway. I could see my mom and dad standing just beyond the fence in front of the car. I waved at them and they waved back when they saw me. My dad appeared to be grinning but my mom looked very serious, even from that distance. As I was cleared onto the runway by the control tower, I saw them put their hands over their ears. The Phantom was very loud even at near idle power. As I lined up on the runway, I looked back over my right shoulder, and I could see them tightly holding their ears. I thought "You haven't heard anything yet." To me, even the F-4's noise level was impressive. Some people jokingly said that it converted jet fuel directly into noise.

The control tower cleared me for a maximum-performance takeoff and climb to flight level one eight zero, which is eighteen thousand feet. I would be there in less than one minute after takeoff. I held the brakes until the throttles were at eighty-five percent of "military" power. (That meant without afterburners.) The F-4's tires would slide on the concrete or the tires would rotate on their rims if I used any more power with the brakes on. I released the brakes and rammed the throttles into full afterburner. The familiar kick in the back of the seat always felt great. I was already at rotate speed and hauled back on the stick. I pulled the nose rapidly up to about seventy degrees of pitch and held it there as I raised the landing gear and flaps. I passed the far end of the twelve-thousand-foot-long runway at about twelve thousand feet of altitude. Fifteen seconds later I rolled the jet upside down and pulled the nose down to level off at eighteen thousand feet. I then pointed the plane west and headed out over the Gulf of Mexico for the rest of my flight.

I returned home after work that evening to see just how impressed my parents were. My dad said that it was great, but my mom didn't say a word. In fact, she almost acted like she was angry with me. It reminded me of when I was a kid and she was mad at me for something I had done that displeased her. (Like going too fast on my bike?)

Later that evening I asked my dad what I had done to make my mother angry. He said that she wasn't angry, just scared. He stated that when I had pulled the nose of my F-4 up and just seemed to disappear into the sky that my mom was petrified.

I have never understood why women appear to be mad at men when they're actually worried about them. Perplexing!

After I started writing these stories, I asked my mother if she remembered watching me take off that day. She got that same look on her face and just simply said, "Yes."

I still don't know if she was impressed.

Go get him

MacDill AFB, Florida, 1964

I don't remember why Dick and I had flown to Ellington AFB, Texas. I'm sure that it was very important business, or we wouldn't have gone. We had spent the night on the base and accomplished whatever it was we had to accomplish during normal duty hours. We went to the club and ate an early dinner before going to operations to file our flight plan and check the weather for our return to MacDill AFB. We needed night flying time, and that is why we were waiting.

The weather was good all along the route. In fact, it was to be one of those Gulf of Mexico nights without a cloud anywhere along the coast. It was going to be a beautiful evening. Sure enough, the sun set "just as forecast," and we took off about thirty minutes later. The moon hadn't come up yet, and it was getting dark quickly.

We left the lights of Houston and Galveston behind us and headed out over the Gulf. We flew southeast along an airway that was almost a straight shot to St. Petersburg. We could soon see the lights of Lake Charles and then Alexandria, Louisiana, farther inland. We were heading toward a point just south of New Orleans, and then there was hardly even a change of heading all the way to St. Pete and home.

We were configured with two three-hundred-and-seventy gallon external fuel tanks and a *little* travel pod on the centerline mount of the fuselage. The pod held our crushed clothes and our frozen shoes, shaving cream, and shorts. It wasn't much, but it was the only place we had to carry anything in a Phantom II.

We were cruising at thirty-seven thousand feet (flight level 370). Our speed was our normal cross-country speed of Mach .9, which is nine-tenths the speed of sound. That gave us a true airspeed of five hundred and forty knots (620 miles per hour).

We had just checked in on the radio with New Orleans and could see the city glowing in the distance. At about the same time, Dick called a radar contact dead ahead of us at about thirty miles. He "locked" the radar onto the return, and it indicated that we were overtaking the other aircraft by about eighty knots. That meant he was cruising at about Mach .82, a very respectable clip.

New Orleans called us at about the same time and said that we were overtaking a National Airlines 727 at our twelve o'clock position for twenty eight miles. He then told us that the National jet was two thousand feet below us at flight level 350 (35,000 feet). We told him that we had radar contact on the airliner, and he "Rogered" us.

We then heard him call the airliner to tell him that he was being overtaken by an Air Force jet fighter. We couldn't hear the airliner's reply because he was using VHF (very-high frequency) radios and we only had UHF (ultra-high) frequencies. (UHF was the common frequency band for the military. VHF was the only frequency band used by the domestic airlines. We could hear the air traffic controller talking to the 727 because he was using a common microphone switch that could transmit on both radios at the same time.)

The air traffic controller then said, "Roger, it's an F-4." Evidently the 727 captain had asked our type when he received the information that he was being overtaken.

A few minutes went by. I could see the navigational lights on the airliner. By now we had closed the distance between us by another couple of miles.

"Jerry, look at that overtake speed," my backseater said over the interphone. I glanced down, and it had dropped off about twenty knots. As I watched it, the overtake speed "gap" on the radar dropped still farther.

"He's trying to keep us from catching him," Dick said.

"It sure looks like it," I replied. I then looked at my fuel gauges and did some mental calculations.

"We can't let him do that, now can we?" I asked over the interphone.

"I just don't see how we could live with that," my backseater said gravely.

"New Orleans, F-4 (Whiskey 91) has a request," I called.

"Go ahead with your request," he answered.

"Request to change our true airspeed from five hundred and forty knots to one zero zero zero knots," I replied. (Remember, the 727 crew could not hear my request.)

The controller, who I know had been watching the airliner accelerate with interest, said, "Go get him!" I suspect strongly that he didn't say that on his VHF radio.

I lit the afterburners, and we began to accelerate rapidly. When we hit one thousand knots true airspeed, I pulled the throttles out of maximum burner and kept enough power to maintain one thousand knots (1,150 miles per hour). The airliner was rapidly growing larger in my windscreen. We were covering over nineteen miles every minute while the 727 was traveling about nine. He was actually going pretty fast, but then almost everything is relative isn't it?

The moon had not come up yet. It was *real* dark. That was *real* good. We were closing rapidly, almost six hundred miles an hour faster than the National Airlines jet.

I pulled the throttles out of afterburner range about a half-mile behind the airliner and "coasted," using our huge overtake speed. I offset my Phantom maybe fifty feet to the left side of the 727, but remained well above it. I wanted to be on the "four striper's" side. When we had moved almost directly overhead the airliner's cockpit, I again selected full afterburners and lit up the night sky with twin streaks of blue and reddish flames.

Although we were two thousand feet above the airliner's altitude, I know it must have gotten their attention. I figured that they probably heard the roar before they saw the flames.

Trying to outrun an F-4! How rude!

Yeah, I knew that it was childish at the time. So?

Now I was happy, except I wished that I could have traded salaries with that airline captain!

How many windows do KC-135s have?

MacDill AFB, Florida, 1964

Air refueling was a new experience for most of us. The largest percentage of the fighter pilots at MacDill Air Force Base had been flying F-84s in the Air National Guard. They had been called to active duty in the Air Force during the Berlin Crisis in 1961. They had brought their unit's aircraft with them to MacDill and had formed the Twelfth and Fifteenth Tactical Fighter Wings. The wing was flying the F-84s while our new F-4s were being built in St. Louis. The F-84 was not air refuelable. Others, such as myself, had recently transferred from the Air Training Command and our training aircraft had not been air refuelable either.

We had flown our first few air-refueling missions with instructors present, but now we were checked out and on our own. This particular mission we were to take off near sunset from MacDill AFB and rendezvous with our KC-135 Tanker at a point over south Florida after dark. We were required to perform a certain number of day and night refueling missions every six months to maintain our combat-ready status.

The flight leader that particular night was a little "weak." Even though he held an important staff position in our wing, he was not a good fighter pilot. He was definitely not a good flight leader either.

Our four-ship of Phantoms took off on schedule and headed south-southeast toward the southern tip of Florida. As we headed toward our rendezvous with the KC-135 tanker, the sun sank into the waters of the Gulf of Mexico on our right. It was the end of a beautiful day and was crowned by a glorious sunset. I have often stated throughout my flying career that folks who have never seen the glories of this world from "on high" just don't realize what they are missing. It's one of the many wonderful perks of the realm of aviation.

As the light levels began to drop, we all adjusted our aircraft lighting accordingly. The sky ahead as we turned toward the east was a deep purple already, and the moon had not yet arisen. This was going to be a "true" night air refueling, not one "logged" just after official sunset while there was still quite a lot of available light. It was going to be an especially dark night.

We could see the lights of Miami swing past the nose as we turned back toward the northwest and our air refueling rendezvous point. Our flight leader had just changed us over to the air-refueling frequency on our radios, and we checked in. The Miami air traffic controllers had relinquished control of our flight to our flight leader, and we were now in what is called MARSA, which meant that the "Military Accepts Responsibility for Separation of Aircraft." Our flight could no longer talk to Miami Center because we were only able to tune one radio channel at a time, but our air refueling tanker could talk to our flight leader and still maintain radio contact with Miami Center. They would maintain this link with Miami Air Traffic Control Center throughout our refueling and would help us obtain our air traffic clearance to resume the remainder of our flight plan after our refueling.

It was really dark now. The tanker pilot had made his left-hand turn back toward our northbound refueling course and was waiting for us to catch up with him. Our flight leader had called a visual contact with the KC-135 tanker. I was flying number four and had little time to look anywhere else but at number three's right wingtip light and the other lights located on the fuselage. I had to maintain the lights in their same relative position to assure that I was in the proper formation location. Number two and three were doing exactly the same thing, so they couldn't look elsewhere either. I felt like we were climbing but thought it was probably just a false sensation. The flight leader gave the signal, and we moved our formation into a right echelon. That meant the lead F-4 was on the left with everyone else in a straight line to the right and angled back about thirty degrees. That was the formation normally used to join on the tanker's right wing.

The tanker boom operator, whose position was in the lower rear of the KC-135 fuselage, indicated that he still did not have a visual contact with our formation. Our flight lead assured him that we were closing toward the right wing of the Stratotanker and he should see us soon.

I was looking along the thirty degree line of F-4 cockpits when I saw our "tanker" beyond them. I was puzzled. There was a long row of lighted windows all along the side of its fuselage. I was just starting to try to sort this puzzle when someone in the formation said, "How many windows does a KC-135 have?"

Suddenly I began to slide forward in relation to lead's plane as did my other two formation mates. I jerked the throttles to idle and extended the speed brakes to maintain my position on number three as he did the same to stay with number two. Two was trying to get back into position with lead. Lead was dropping like a rock to descend back into our air refueling altitude block where we belonged.

I don't think anyone on the Eastern Airlines DC-8 saw us. At least I never heard about it if they did. After we located the KC-135 and completed our refueling, Miami Center did ask us if we "enjoyed our air refueling." I had never heard an Air Route Traffic Control Center say anything like that before, and I haven't since. The KC-135 crew could hardly do their job for the "snickering."

I would have loved to have heard the conversation between Miami and our tanker while our flight leader was taking us on our little excursion.

I told you at the beginning of this story that he was "weak." And that's being kind.

One tough bird

MacDill AFB, Florida, 1964

McDonnell builds a strong airplane. A friend of mine, Willy Barker, who now lives in Fayetteville, Arkansas, can attest to the accuracy of that statement.

One of the roles of the F-4 was nuclear weapons delivery. To perfect it, we used a bombing range near Avon Park, Florida. A portion of the range was used exclusively for simulated nuclear deployment. It consisted of a "run-in line" on the ground, several thousand feet in length, with discernible points along it called IPs, or initial points. Utilizing these points, with the help of timers in the cockpit, an F-4 crew could know precisely when to begin a pull-up for a bomb toss. The simulated weapon would then automatically release along a precalculated path to the target. Very large concentric circles around a center point, or bullseye, would aid in the scoring. A marking charge in the fake bomb would go off on impact and an observer in a range tower would score the accuracy of the delivery.

A nuclear practice mission would begin with a low-level flight between one hundred to five hundred feet above the ground for a hundred miles or more. The speed at which we normally flew these missions was four hundred and eighty knots, about five hundred and fifty miles per hour. The flight controls on the Phantom II were extremely sensitive. When I flew it at low altitude and high speeds, I would press my right forearm into the top of my right leg to keep from moving the stick unintentionally. A movement of fractions of an inch could result in quite large pitch excursions and G forces. The airplane had built-in stability augmentation systems to assist the pilot in flying it. For those of you familiar with yaw damping systems on some aircraft, it was similar, but in all three axis instead of just the rudder. If the "stab aug" was inoperative, we were not supposed to take off. Without the system operational, the F-4 was very difficult to fly, especially at high airspeeds at low altitude.

My friend Willy had just completed the low-level portion of his flight and was now flying down the nuclear run-in line. Being a few seconds late at the first IP, the frontseater pushed up the throttles in the Phantom into the afterburner range. The plane surged ahead, accelerating rapidly. At about this same time, the pitch stability augmentation evidently failed. The F-4's nose pitched upward as the afterburners were deselected. The frontseat pilot corrected. The nose pitched violently downward. The pilots were thrown upward in their seats. As that happened, the pilot's hand on the stick was pulled backward. The nose pitched violently upward again. How violently? The observer in the scorer's tower said the nose pitched as much as forty-five degrees up and down, at least four times, in a distance across the ground of two to three thousand feet. With the Phantom traveling now at about six hundred miles an hour, it traversed that distance in approximately three seconds.

Willy, in the back seat for this flight, tried to eject. The G forces were so high that he couldn't raise the ejection handle located between his legs. He felt the airplane snap roll and was buried in his seat by the G forces. Suddenly he realized that he was still alive and the airplane was in a climb. He let go of the ejection handle as the aircraft commander regained control of the Phantom II. There were warning lights everywhere in the cockpit, but they were still airborne. MacDill was just a few miles to the west, and they made it back, flying at a relatively slow speed.

As an indication of the forces involved, both engines had been torn from their mounts and were lying in the bottom of the engine bays. They had tilted to such an angle that the exhaust was burning the paint on the tail. The left three-hundred-and-seventy-gallon fuel tank attached beneath the wing was torn completely from the jet and had knocked the top six inches off of the vertical stabilizer.

As far as I know, that airplane was repaired and flown again. It was estimated that they had pulled about thirteen positive Gs, and seven or eight negative Gs. That is a phenomenal accomplishment for an aircraft. Especially when the final positive Gs were pulled during a snap roll at probably three hundred feet of altitude and more than six hundred miles an hour. Rolling G forces are much more destructive to an aircraft than with the wings at a constant angle of bank.

About ten years later, my Fayetteville friend suffered some delayed results of that day. He and I were members of the same Air Guard unit flying RF-101 VooDoos. One day, while getting out of his aircraft, he felt a debilitating pain in his back. A calcium spur had penetrated a nerve. He consequently had to have a serious back operation. Of course, it can't ever be proven, but I think that it was an outcome of that wild ride in Florida.

What do you mean you *fell* out?

MacDill AFB, Florida, 1965

The skills of air-to-air combat need constant honing. They are also some of the most difficult talents to master. In an aircraft like the Phantom II, learning to fly the aircraft to its maximum potential was eminently demanding. It had such tremendous performance capabilities that it was sometimes an accomplishment just to stay with it. One sunny day over the Gulf of Mexico, another good friend of mine didn't.

Maxey was the backseater of one of my squadron's F-4s engaged in a "furball" with several others from our unit. It was a clear day, and the sky was filled in the practice area with twisting, rolling, and roaring Phantoms. Their radar contacts had rapidly become visual contacts with the adversaries of the day closing at a combined speed of over fifteen hundred miles per hour. The fight was on. Maxey's aircraft commander pulled up into a rolling vertical scissors maneuver with another F-4. They were running out of airspeed and were canopy to canopy with the other Phantom. Maxey's frontseat pilot shoved forward on the control stick to get separation from the other fighter and to accelerate.

Maxey looked up and saw his parachute deployed. The F-4, in which he had just been sitting a few moments before, was twisting and diving away from him in full afterburner. One of his shoulders hurt like crazy, and he could hardly move the arm on that side. Unknown to him, his aircraft commander was trying to talk to him on the interphone in the cockpit. The commander had suddenly heard a loud noise, and air was rushing past; he didn't know what had happened, but he did know Maxey wasn't answering. Everything up front looked normal, and he still couldn't figure it out. Then he saw Maxey. Maxey was hanging under his parachute and slowly drifting toward the blue water below.

As the F-4s circled the area, they could see that Maxey was alive. One of them radioed for search and rescue aircraft. They stayed with the downed pilot as long as they could before their fuel state forced them home. Maxey had a tough time getting into his raft because that sore shoulder was dislocated. He spent several hours in the water before the rescue helicopter reached him; he was okay, except for the shoulder.

Everyone, especially his frontseater, was wondering why Maxey had suddenly elected to exit his F-4. It turns out that he hadn't decided to bail out. Whoever hadn't bolted Maxey's ejection seat back into the aircraft after maintenance had decided for him. Seems that when the aircraft commander shoved the stick forward, the ensuing negative Gs started the unsecured ejection seat sliding up the seat rails. When it reached a certain point, it fired the canopy jettison mechanism, but the ejection seat itself never fired. It just continued up the rails until it was at the top. It then tilted backward, with Maxey still attached, and whizzed along the fuselage until it bounced off the vertical stabilizer of the Phantom. This was surmised because Maxey had paint on his helmet from the red, blue, and yellow Tactical Air Command insignia on the aircraft's vertical tail. Colliding with the stabilizer is more than likely what injured his shoulder.

To this day, Maxey is the only one I have ever known to fall out of an F-4, and I still kid him about it whenever I see him. Can you imagine bouncing your head off the tail of an aircraft going several hundred miles an hour and living to laugh about it?

Maxey was a high ranking officer in the Mississippi Air National Guard the last time I saw him. As far as I know, he hasn't fallen out of any more airplanes.

Air defense of the what?

Naha Air Base, Okinawa, 1965

Things were heating up in Southeast Asia. It had been decided in early 1965 that the squadron of F-102 Delta Daggers based in Okinawa should be transferred to Don Muang Airport at Bangkok, Thailand. I suppose that because the Thai government officials were allowing the United States to stage missions from their country into North and South Vietnam, they felt their Air Defense should be expanded. As a result, the Air Defense role of Okinawa, or the Ruyuku Islands, as they were called, became a mission of our F-4 wing at MacDill.

Every three months, a different squadron of the Twelfth Tactical Wing would arrive to take over the role. We were the second squadron to arrive at Naha Air Base. It was about the middle of March. We were flown to Okinawa in Air Force cargo jets and were to crew the F-4s that had been flown there three months earlier by our sister squadron, whom we were relieving. We would fly these F-4s for the three months of our temporary duty, then ferry them back to MacDill.

Our ninety days in Okinawa proved to be a sobering and maturing process. We had departed the world of training and preparation and entered the operational world. The proximity of Red China, and our somewhat tenuous political relations with them at the time, kept our minds and attitudes on a more serious track than they might have been at another time or location.

The loss of two of our squadron mates in a tragic night training accident soon after our arrival also served to remind us that we were in a dangerous profession.

Fighter pilots ain't grown men!

Naha Air Base, Okinawa, 1965

Naha Air Base had a nice Officer's Club. It served very good food, and because it was one of the few places to eat on base, it was very popular.

A contingent of C-130s was permanently based there and our squadron was there on TDY (temporary duty). The permanent party of course had their dependents with them. We had to be there for about three months without our wives and children.

Several families were scattered around the club that evening having dinner. I was sitting at a table with my backseater and two other "quieter" types like ourselves. (Believe it or not, there are quite a few fighter pilots who *don't* drink and chase women. Really!) We had just finished our main course and were having coffee and dessert. It was very quiet. Suddenly, on the other side of a large curtain that hung all along one side of the dining room, a piano started playing. It was just medium volume and not that annoying. In fact, it was some kind of honky-tonk song and started to lift our mood. When you are separated from your family for months at a time, it can be somewhat depressing.

Now another sound was added to the piano music. It was a kind of scraping, shuffling noise. We were wondering about the new sound when a third sound joined in. There was no mistaking what it was. It was a woman screaming at the top of her lungs! Almost in unison we jumped up and headed for the sounds. The piano and the shuffling noise stopped. As we cleared the curtain ready to rescue this lady in distress, we stopped in our tracks.

On top of the beautiful black baby grand piano was one of our squadron pilots. He was still dressed in his flight suit and boots. One of our other pilots was sitting at the piano with his hands still poised over the keys.

"Lady, what's the matter?" yelled the startled two-hundred pounder on the piano.

"Grown men don't act like this!" she screamed.

He seemed to think that over for a few seconds and then said, "Fighter pilots ain't grown men!" The piano player started up again and the dancer resumed his shuffle.

It took several of us to convince the "entertainers" that they needed to leave quietly and immediately. By that time a small crowd had gathered. There was no question about the two pilot's organization. They had our squadron patches all over their flight suits, and it was a very small base. We decided it was time for us to make our exit also.

The next morning at the end of our daily squadron briefing in the flight operations building, our squadron commander stood up. It seems that the screaming lady had been none other than the Base Commander's wife. He was a "full" colonel, and for those of you who aren't familiar with such matters, an Air Force officer's wife is understood to always be one rank higher than her husband. (At least at home!) Our lieutenant colonel commander then proceeded to tell us approximately how many pounds of his butt he had just had chewed off that morning by her husband, the "full" colonel.

He continued after a spate of snide remarks about needing to lose a few pounds anyway: "Now the rest of you are going to get a lot lighter. The operations officer is going to pass among you for your contributions toward refinishing the Officer's Club piano. He will continue to pass among you until the full amount is collected. We are not leaving this room until we have it."

I can tell you that the refinish job was very expensive! I thought at the time that they could have bought a new piano for that much. Maybe they did.

And that remark about "Fighter pilots ain't grown men"? That's one that I'll have to agree with, in some cases.

Needless nightmare

Naha Air Base, Okinawa, 1965

It had been raining for three days. It wasn't just light rain or showers, but a hard, steady downpour. There was a typhoon somewhere between us and the Philippines. We kept four F-4 Phantoms on alert twenty-four hours a day, seven days a week. Two of these aircraft were expected to be airborne within five minutes of a scramble signal from a loud klaxon horn located within the alert facility. The horn was controlled from a command post that was also manned twenty-four hours a day. The other two F-4s were on fifteen-minute alert if the situation warranted their launch. The command post was in direct communication with various defense agencies and the region's air traffic control system.

There were several levels of alert to which we adhered. These not only were affected by military or political situations, but also by weather. If the weather was so bad that it was extremely risky to fly, the alert status was called "Mandatory." It was raining so hard that we had been on Mandatory for two days. This meant that a tactical situation had to appear extremely serious to the command post controller before he would launch us into the terrible weather conditions. Also affecting our flying locale was the fact that there was no other airport to which we could divert except Kadena Air Base, just a few miles away on the same island. In fact, Kadena was so close that it usually experienced the same weather conditions as Naha Air Base.

Four crews, each consisting of a frontseater or aircraft commander, and a backseater co-pilot served a twenty-four-hour tour of alert. At the end of twenty-four hours, we would be replaced by four more crews from our squadron. I and my backseater had been assigned as one of the fifteen-minute crews when we had reported for duty at seven that morning. At 7 P.M., we would become the leader of the two five-minute alert aircraft. By dividing the five-minute alert duties, the tension that accompanied the knowledge that you had to be airborne within five minutes was spread among the pilots. After the dining hall had brought our evening meal and we had eaten, we were all engaged in various activities to pass the time. There was ping pong and a shuffleboard game and a small library of books and magazines. Things were fairly relaxed in light of the fact that a launch into this weather was a highly unlikely possibility. In fact, that had been the main topic of our dinner conversation. To everyone's knowledge, no one had ever been launched from Okinawa under such conditions. It was just too dangerous. If you couldn't get back into Okinawa, there was absolutely nowhere else to go.

The sleeping area of the alert facility was kept dimly lit with red light. This was to prevent the temporary loss of our night vision in case of a night launch. For the same reason, the lights in the alert hangar where our Phantoms were kept and most of the lighting in the cockpits were red. For the first time, I decided to take off my flight suit when I went to bed that night. Always before when I was on five minute, I had slept in my flight suit just in case. But we were on Mandatory, right? The red light glowed dimly. The rain on the roof was hard and steady. Its sound lulled me to sleep.

The klaxon drowned out every other sound, including the rain! It was like a bad dream; but as I jerked awake, I realized it was no dream. Someone, or something, had really set off the klaxon horn. I almost stepped on my backseater's head as I rolled out of my top bunk and began struggling with my flight suit and boots. Finally my "quick don" flying boots were zipped and I headed into the alert hangar on a dead run. As I settled into the front seat of my F-4, the big hangar doors started up. The rain coming down looked like someone was pouring it out of a cattle trough on the roof. I couldn't see more than a few feet into it.

As I started the engines, I was hopefully watching the door into the alert facility. One of the fifteen-minute alert pilot's additional duties was to call the command post on a direct telephone line in case of a klaxon. He would confirm that it was real and not a mistake or an electrical malfunction of some sort. I knew that this had to be an error! They wouldn't launch us into what I was seeing just beyond the hangar door now fully open. I knew that the confirming pilot would soon be running to the door and giving us a "finger across the throat" signal, meaning shut down the engines and stop the launch. Sure enough, there he was, but he was giving us a "circle above the head" signal. That meant "launch!"

My crew chief checked my connections, pulled my ejection-seat safety pin out, and scurried down the ladder. As he headed to safety, I closed my canopy and my backseater did the same. He was also commenting on what he thought of our current situation, but I won't go into his descriptive and colorful terms. I will say, however, that I didn't disagree with him. As I added power and taxied out of the alert barn, my wingman checked in. I called the tower, "Whiskey One, launch two."

Hoping to hear the control tower cancel our launch, I heard instead, "Whiskey cleared to launch, vector three two zero, gate climb, angels forty, V-max." The translation was for us to turn to a heading of three hundred and twenty degrees after takeoff, climb with full afterburner power, level off at forty thousand feet, and then go as fast as we could.

After finally locating the runway, which was only a few yards from the front of the alert barn, I selected the afterburners and the Phantom leaped forward. I had never seen it rain so hard. Three and one-half minutes after being awakened from a sound sleep, there I was trying to focus on the attitude and airspeed indicators as I raised the landing gear handle. Our standard procedure was to fly runway heading for forty-five seconds after liftoff before turning to the assigned heading. I realized that I had forgotten to start my timing. I figured that at least forty-five seconds had passed, and I turned left to a heading of three hundred and twenty degrees. My adrenalin had me fully awake now, and I pegged the airspeed indicator at three hundred and fifty knots. At our aircraft weight, and configuration, we were climbing at around twenty to twenty-five thousand feet per minute.

Suddenly in my right peripheral vision, I saw something in the clouds. My wingman's red left wingtip light was coming right at us. I shoved forward on the control stick, and he passed right over the top of the canopy! I turned five degrees right and told him to hold his present heading. He "Rogered" me, and just then we popped out of the top of the dense clouds at thirty-five thousand feet. Whiskey Two was about a half mile to my left as we both rolled inverted and pulled our fighters' noses down to level off at our assigned altitude of forty thousand feet. My backseater was still commenting about our present state of affairs, especially after our near-midair collision seconds before. I still agreed with him.

As we accelerated to V-max (1.6 times the speed of sound, with external fuel tanks attached), I checked in with the GCI unit. Those initials stand for "ground-controlled intercept." It was a large ground-based radar unit whose job was to vector fighters to intercept possible airborne targets.

GCI answered, "Possible bogey, eighty miles, twelve o'clock, very high." That meant that we had a possible hostile aircraft directly ahead of us at eighty nautical miles, and well above our present altitude of forty thousand feet.

My backseater found the target with our radar and called "Whiskey One, contact."

The GCI controller "Rogered" and said, "The target appears to be turning toward the south."

Sure enough, the "blip" on our radar screen began a drift toward the left. I began a left turn to cut off the target. My backseater said over the interphone to me, "Jerry, that guy's above sixty thousand feet according to my calculations." He had just locked on to the target with our radar and the equipment began giving us more information.

Just then the GCI controller said, "Whiskey flight, we request that you get a visual identification on the target."

I immediately responded, "Roger, request permission to drop tanks." Our top speed with external fuel tanks attached was Mach 1.6. We could tell that this target was going very fast, and we'd need all the speed we could muster to get up to him. Mach 1.6 was just over one thousand miles per hour, and we were already going almost that fast.

The GCI controller finally answered, "Permission to drop denied."

About that time we rolled out behind the target, and it all became a moot point anyway. Our radar was giving us the information that this guy was going faster than our top speed without tanks. Not only that, he was above seventy thousand feet. Our maximum sustained altitude was around sixty-six thousand feet. We were carrying missiles that might have shot him down earlier in the intercept, but there was no way we were going to visually acquire him that night!

I jerked the engines out of afterburner as I remembered my real problem. I was half afraid to look at the fuel gauges. We had been airborne for only twelve or thirteen minutes, but we'd been operating in full afterburner the whole time. The Phantom was not built for its fuel efficiency. It was a gas hog even without afterburner. I turned rapidly back toward Okinawa and called for a fuel check from Whiskey Two. Thank God, he had a little more than I did, but we were definitely in trouble. We were one hundred and eighty miles from Naha Air Base.

My backseater began furiously figuring our probable fuel remaining when we arrived there while I began trying to milk every mile out of the F-4 that I could. I checked in with the air traffic control unit at Okinawa and mentally crossed my fingers as I asked them for the weather at both Naha and Kadena air bases. The miracle had not happened. Both air bases were zero ceiling and zero visibility. My backseater then broke more good news to me. We only had enough fuel for one approach and landing. If we missed it, there should be just enough left to pull up and eject over the beach.

Under normal conditions, we were not even supposed to attempt an approach and landing when the ceiling was less than two hundred feet or the visibility less than one-half mile. These were not normal conditions. It was 3:25 A.M., we were almost out of fuel, and we had no place else to go. I flew the most precise radar approach I'd ever flown. The ground-controlled radar approach technician was calm and deliberate. He knew this had to be one of the best approaches he had ever controlled or we might be history. The rain was slamming off the windscreen. Even though I had turned on the windscreen rain removal system full blast, it was doing nothing. We arrived at the point in the approach where the controller usually told us, "Take over visually and land." Our controller kept talking. He knew we couldn't see anything.

I descended right through the usual minimum descent altitude of two hundred feet. I kept going down another one-hundred-and-fifty feet. Finally, I gave up and started to add power to go around. The wheels touched! I yanked the throttles back to idle and grabbed the drag-chute handle just as a white runway light rushed by in the murk. It was too close to us and I began steering the Phantom away from the right side of the runway. Another runway light went by, a little farther away this time.

I stepped on the brake pedals, and the antiskid began to cycle furiously. Gradually, the F-4 began to slow down. I knew I had to get off the runway quickly because my wingman was right behind me. In fact, I could hear him talking to the approach controller about the existing conditions. I interrupted and told him exactly what had happened, and that I was looking for a taxiway to get off the runway. I then reassured him that I would be clear. Now I had to do it! He only had one chance, too.

There . . . a blue taxiway light was just ahead to the right. I taxied just past it and turned. I wasn't in mud, so I figured I must be on concrete. I taxied far enough to ensure that my drag chute was clear of the runway and then stopped. Now I began worrying about my wingman. I could hear the controller talking him down, exactly as he had just done for me.

Finally the controller said "You're now over the touchdown zone, on centerline."

Whiskey Two yelled, "We're down!"

I called him and told him, "Just stop it on the runway, and don't try to turn off." I said I didn't know exactly where I was and nobody else was dumb enough to be up flying around in this stuff anyway!

I called the tower and told it to send a tug out to find us and tow us in. They asked where we were. I told them if I knew that, I would have taxied in. As I shut the engines down, I was thinking how close we had just come to having to eject. I reached up to my shoulders to disconnect the two fittings that attached me to the aircraft ejection seat with its built-in parachute.

My legs started to shake as the shock set in. The fitting at my right shoulder wasn't connected. My parachute would not have opened.

The rain didn't let up for two more days.

Postscript: For those of you who wonder, the possible bogey was ours. Its flight plan had been misplaced.

Its nickname was *Blackbird.*

Like a farmer, outstanding in his field

Naha Air Base, Okinawa, 1965

My backseat pilot in Okinawa was a great guy. (I'm sure he still is, although I haven't seen him in years.) He had been a radar observer in the Air Defense Command before getting a pilot training slot. He certainly didn't intend to wind up right back in the rear seat of an airplane directing intercepts again. He used to refer to himself as a "PSO," which stood for "Plenty-Screwed Officer." I'll refer to him as Dick, his real first name. PSO also happened to be the original official title for F-4 backseaters, which was Pilot, Systems Operator.

He didn't let his "screwing" affect his attitude or job performance. He was an excellent pilot, and I let him fly the jet as much as possible. He was absolutely the best at directing intercepts that I had ever seen. In fact, he was such an expert that our flight commander recognized it and wanted him to frequently brief the other members of the flight on air-to-air intercepts.

Our flight commander was a major at the time and a very sharp guy. He was dedicated and hard working. It was obvious to me at the time that he would go far in the Air Force. He retired as a major general! However, the flight commander had one very annoying habit that just absolutely drove my backseater nuts. While Dick was trying to give the very briefings that the commander had asked him to give, the flight commander would continually interrupt to give his own ideas and suggestions. This completely messed up Dick's well-planned and precise briefings, and besides, the commander's inputs were not always correct. This continued until one day Dick's face turned red, and he just threw up his hands and walked out. I don't think the flight commander had a clue as to why. He just didn't realize how disruptive and irritating he was being.

Several days later, Dick and I were in the base exchange. He walked over to the greeting card section to find a card for one of his family members back home. Suddenly he reached over and picked up a card. He had a very mischievous look on his face. He handed me the card and grinned. I looked at it. It had a large block-letter message on the front that read "TO A MAN WHO IS OUTSTANDING IN HIS FIELD!" I opened it up, and there was this cartoon character of a farmer standing out in the middle of a hay field with a piece of straw in his teeth and a silly grin. The writing said, "LIKE THIS FARMER!"

I looked at Dick and asked who it was for.

He said, "You'll see."

Dick couldn't wait to get the card in the mail. He addressed it and mailed it from the local post office. He then began watching our commander like a hawk. We had our mail delivered to our squadron area, and sure enough in a couple of days it came. We watched as the flight commander turned it over and looked for a return address. There was none of course. He opened it up and looked very pleased when he saw the message on the front of the card. Then he opened it to the inside. His countenance went very dark, and he immediately began looking around the room. We were pretending to be very engrossed in our duties at the time until he left.

Dick burst out laughing and said, "Did you see his face? Man that was great!" But he wasn't finished. Our flight commander was scheduled to fly, and as soon as he was gone, Dick went over by his desk and checked the wastebasket. The card was there. He picked it up and put it in his pocket. I asked him what he was going to do with it and he said, "You'll see."

After several days Dick began hanging around when the mail was delivered. Again he watched the commander closely. Finally! The flight commander looked at the envelope. Again there was no return address, but it was mailed from Tampa, Florida, where our home base of MacDill was located. He opened it, and a look of disbelief came over his face. He again started checking everyone's face in the room to see if he could locate the culprit, if he were one of us. Again we were the picture of innocent activity. He disappeared.

The next day Dick came walking up to me and said, "Look what I've got." It was either the card, or one just like it.

"Where did you get that?" I said.

"I found it under his desk blotter," he replied.

Weeks passed and Dick went into his flight-commander-observation routine. He didn't want to miss it. Again the mysterious card came back. This time the envelope was postmarked in Germany!

I thought both the flight commander and Dick were going to "blow a gasket," Dick from trying to keep from laughing out loud. The furious flight commander stalked out, and I finally breathed.

Dick found the card in the commander's wastebasket again. This time it was shredded in what looked like a hundred pieces. I think he thought about trying to glue it back together but decided against it.

I know Dick actually felt like I did about our flight commander's abilities—that he *really was* "OUTSTANDING IN HIS FIELD."

General "T,"

Best wishes.

I hope you are doing well. *Now* you know!

General "C"

P.S. Dick, I hope you'll forgive me for "spilling the beans." Or was it hay? I think you exhibited great wisdom in quitting while you were ahead. Jerry

Say your type aircraft

Naha Air Base, Okinawa, 1965

It was tough. The F-105 "Thunderchief" drivers were merciless.

"What's wrong with your two-man bomber?"

"Why is your Mach II, Phantom II, just sitting there and 'pissing' on our ramp?"

"Man, get that ugly thing out of here. It's scaring my kids!"

These were just a few examples of the verbal "abuse" we had been taking the last couple of days. We had been sent to an F-105 base in Japan for what reason I don't remember, but I sure was regretting it. We had spent the night. The next morning we had received the unwelcome news from Transient Maintenance. We had a bad hydraulic leak. It was going to take a couple of days to fix. All we had brought to wear was our flight suits, fresh socks, and underwear because we were only supposed to be there overnight. We had to travel light because there was no place to carry personal items in an F-4.

The problem was not so much having just flight suits. It was the "Phantom Phlyer," "Mach II Club," and other various F-4 and squadron patches on them. We caught a lot of grief about our patches from the local F-105 fighter pilots, but we weren't about to take them off. Every fighter pilot is proud of his particular machine. Their banter was all in good fun, but I sure would be glad when our jet was fixed.

Finally it was. It had been sitting there forlorn looking with maintenance guys working on it for a couple of days while the Thunderchief pilots taxied by and laughed and made "rude" hand signals. We started engines and returned several hand signals to some F-105 jocks as we taxied for takeoff. At the end of the runway, I asked the tower for permission for a maximum-performance climb to our cruise altitude of thirty-one thousand feet. (At this point, it should be understood that the F-105 was not an exceptionally sterling performer right after takeoff. It was about the same weight as an F-4, but it had a single engine with eight thousand pounds or so less thrust than the total of our two J-79 engines.)

The tower approved the climb and cleared us for takeoff. I taxied onto the runway and held the brakes as I increased the power to eighty-five percent. As we started the takeoff roll there were several Thunderchiefs waiting to taxi onto the runway. I held the nose of the Phantom down after we were airborne and raised the landing gear and flaps. We were maybe ten feet in the air and doing almost three hundred knots as we passed the end of the runway. I hauled back on the stick. It was "pay back" time. We shot nearly straight up from the runway's end. I could see the air base in my mirrors as it dropped rapidly away.

"Whiskey, contact departure control," said the tower.

"Yokota departure, this is Whiskey Zero One," I called.

"Whiskey, call passing niner thousand feet," requested the controller.

"We're past niner thousand," I said.

"Roger, call passing one five thousand," came his request.

"We're past one five thousand," I answered.

There was silence for a few seconds as the controller checked his radar.

"Okay Whiskey, call passing two five thousand," he finally requested.

"Whiskey Zero One has passed twenty-five thousand," I gleefully said.

"Whiskey Zero One, say your present altitude," came the exasperated controller's request.

"We're level at thirty-one thousand," I said.

"But Whiskey, my radar only shows you about nine miles from the Tacan," came the puzzled-sounding controller's call.

"Yeah, and six miles of that is *altitude*!" my exuberant backseater answered.

"Say your type aircraft, Whiskey," the controller's voice requested. The F-4 was still pretty new, and the controller was used to the much slower climbing F-105s.

"That's classified," came my backseater's "smart-assed" *and* incorrect call, but he was having a ball like I was, "giving it back" to the F-105 folks.

"Roger Whiskey. *Understand* you're an *F-4.* I knew you weren't an F-105 when you told me you were already out of niner thousand while ago. You're cleared on course. Have a good flight," the controller said and passed us off to the en route control.

It had *almost* made the last couple of days worthwhile.

I didn't blame the F-105 pilots one bit for giving us hell. I would have done the same thing. It's all part of the "job."

Missiles are supposed to be deadly! (for the other guy)

Naha Air Base, Okinawa, 1965

The word "Sparrow" does not normally evoke any thoughts of fear. I don't know who named the AIM-7 missiles that we carried half buried in the belly of our F-4s. One wonders why they didn't come up with a more fearsome sounding name like Hawk or Predator, or something aggressive like that. Sparrow? Sounds kind of meek, doesn't it?

The Sparrow was a radar-guided air-to-air missile that was designed to follow a radar beam to an airborne target. It didn't have to actually hit the target to destroy it. It utilized a high-explosive warhead wrapped with a device that created a deadly wide area of destruction when it was detonated. It only had to pass near a target to be activated by a proximity fuse. In this case, the old saying about a "miss being as good as a mile" did not apply.

We flew practice scrambles and intercepts on a regular basis out of Naha. Sometimes they were flown in aircraft equipped for alert with real live missiles. Our four Sparrow and four Sidewinder heat-seeker missiles were each capable of downing an aircraft of any size, so we were extremely cautious when flying with this lethal array.

My targets for the day would be fairly demanding. One was to be flying very low to the sea and the other was to be moving at almost Mach 2 at high altitude. The "scramble" had gone well for myself and my wingman. We were being vectored toward the northwest at around fifteen thousand feet and just below the speed of sound. My backseater found the low-altitude target, and we had "eliminated" him relatively easily during a head-on pass. It was a T-33 flying toward the islands. Almost immediately the radar controller reported a high-speed, high-altitude target one hundred miles north of our position heading south.

I lit the afterburners and began a climb while turning toward the direction of our target. My backseater began his radar search. I initiated a one-half G unload maneuver at thirty-five thousand feet. It was in a half-G condition that the F-4 had been found to accelerate the quickest. As we passed Mach 1.4, the engine vari-ramps opened, and the Phantom began accelerating at a faster rate. The vari-ramps were devices located inboard of each engine intake. At about 1.4 times the speed of sound, they would rapidly open, creating a shock wave across the lip of the engine intakes. The air entering the intakes was effectively slowed somehow by this wave, and the engines developed more power utilizing the affected airflow. It was very effective. I can compare it to stepping down on an automobile accelerator pedal to select passing gear. It was easy to tell when it occurred.

Shortly thereafter, I felt a slight bump, and the right-engine overheat light illuminated. I pulled that engine out of the afterburner range and began retarding the throttle toward the idle-power position. The light went out. The exhaust gas temperature indicated a normal range. I left the throttle in idle after testing the fire-warning system, as prescribed by the checklist.

Since this was only a training mission, I called off the intercept and asked my wingman to join up and look my Phantom over as I turned back toward our base at Naha. My wingman joined up as requested and spent several minutes underneath and behind my jet, checking for anything unusual that he could see. Finally, he joined up on my right wing and reported that everything looked normal to him. I relaxed somewhat, as I had already accomplished all the required procedures based on his visual inspection.

Because I had experienced an overheat light on an engine and had pulled its throttle back and left it in idle, I was essentially performing a single-engine approach. Responding to that situation, the base deployed the fire and rescue equipment beside the runway. After I landed, the equipment followed our aircraft down the runway and onto the taxiway. At that point, I stopped the plane, and a fireman ducked underneath to make a quick visual inspection before clearing us to taxi back to the parking area.

Suddenly, the fireman ran out from under the plane and frantically gave me the engines cutoff signal. I immediately complied, as he excitedly told the others through his hand-held radio what was transpiring. He then began giving us hand signals to get out of the airplane immediately. I was puzzled, but we exited the cockpits as fast as we could. He then grabbed us by the arms and began to run away from the F-4. We jumped into a maintenance van, and it sped away as the fire trucks began backing away from the Phantom. The driver seemed to be in an awful hurry. I asked him why, and he said he wanted to get as far away as possible in case the missile went off.

I said, "What missile?"

He answered, "That Sparrow that your right engine was cooking!"

The Explosive Ordinance Disposal Officer came to see me later that day to brief me. He said that a stainless steel retaining cap had blown off the bottom of my right engine. The engines were basically interchangeable. Bleed air from near the turbine section of each engine was piped out to the wings to be used for what was called boundary-layer control air. It was blown across the wings and flaps and allowed the F-4 to land at much slower airspeeds than would be possible otherwise. Each engine had two outlets for this air. Depending on which side of the aircraft the engine was being used, one of these outlet holes was capped.

My right engine cap had blown directly through the bottom of the engine bay leaving a four- or five-inch hole lined up almost perfectly with the warhead section of my right-rear Sparrow missile. That air was extremely hot air. It was around five hundred degrees Celsius to be exact. It had heated the high explosives in the missile to the point where the white paint on the missile had turned brown. It seems that high warhead temperatures could cause the explosive compounds to become very unstable. A slight jolt or bump could have set it off. That's why everyone had wanted to get the hell away as soon as possible.

He continued, "We have to let the warhead cool for several hours. Then we'll download the missile and transport it over to the ordnance disposal area and get rid of it. I just wanted to let you know what was going on and to congratulate you on your smooth landing!"

I glared at my wingman and his backseater who were sitting nearby. They didn't look at me as I made a caustic comment about their assistance and suggested that they go get their eyes checked. My backseater and I were *not* happy.

Incidentally, those guys were not members of my squadron. They were what we called "division wienies" who came to fly with us on occasion.

The point is probably obvious. Had I known about the hole blown in the bottom of my airplane, I would have shut down that engine and the missile wouldn't have continued getting cooked. Also, I could have ensured that I made a smooth landing by adding a little airspeed and easing the Phantom onto the runway.

Thanks a lot, wienies!

In the line of duty

Naha Air Base, Okinawa, 1965

Kadena Air Base, Okinawa, was home for a wing of F-105 Thunderchiefs, as mentioned in an earlier story. The F-105 was a single-seat fighter-bomber that was beginning to be used extensively in Vietnam. It was extremely fast at low altitudes. I think its airspeed limit was about eight hundred and fifty knots on the deck. It would also carry a big load of conventional bombs and had a built-in twenty-millimeter Gatling gun. It was to carry the brunt of the air war in North Vietnam for some time to come, especially the air-to-ground bombing role in some of the most heavily defended areas of the country.

Two of these jets took off in formation one afternoon and headed toward Southeast Asia. They were unarmed. Their routing took them west by southwest toward Taipei before they were to turn south. That direction also headed them toward Red China. The Thunderchiefs were cruising at about six hundred statute miles per hour when they passed their first checkpoint.

The number two F-105 moved in closer to the leader. They had continued straight ahead toward China instead of turning. His radio calls to the leader were going unanswered. He looked closely at the leader's cockpit. The pilot's head was slumped forward and there was no discernible movement. The Thunderchiefs were boring straight toward the Chinese mainland at ten miles per minute. Number two made an emergency radio call even as he continued to try to get the lead pilot's attention.

Hypoxia is a very insidious disorder. It comes on very slowly and initially causes a sense of euphoria and well-being. The lack of oxygen then begins to affect motor skills and vision. Drowsiness occurs and then finally, if oxygen is not administered immediately, unconsciousness.

Whiskey One was scrambled immediately with his wingman and they roared after the F-105 flight in full afterburner. It was going to be close. The number two F-105 pilot knew that the Chinese radars were probably tracking them by now.

My squadron mate in the frontseat of Whiskey One was being briefed on the situation by the radar intercept controller. As the realization of what his role in this rapidly developing scenario was to be, he felt sick to his stomach. Whiskey One flight was now going as fast as the Phantoms would fly. They could do no more but sit and wait, at nearly twice the speed of sound.

The second F-105 pilot was trying everything he could think of. Desperately, he brought his left wing up closer and closer under the right wing of the leader's plane. As he tried to force the leader's wing upward with the airflow over his, the autopilot of the lead aircraft fought to keep the wings level. Again and again, the second Thunderchief's pilot tried to force the lead F-105 to turn. The F-4s were in radio communication with the second F-105 now. They were also in radar contact and closing the distance between them as swiftly as possible, but it had begun to appear like an international incident in the making. The lead F-4 pilot's emotions were in turmoil. He wanted to catch the F-105, but he also didn't want to catch it.

The F-105 wingman knew time was getting extremely short. He tried one more time to turn the unconscious pilot's Thunderchief with his wing. Its right wing slowly raised and the aircraft began to turn slightly. Suddenly, the lead F-105's autopilot disconnected and the plane rolled over on its wing tip. The wingman managed to wrest his jet out of the way in time.

The unpiloted F-105 continued wing down, lower and lower toward the East China Sea. All the way down, the wingman followed and pleaded for his leader to wake up. He never did.

The F-4s of Whiskey flight safetied their missile switches and slowly turned back toward the Ruyuku Islands. There would be no confrontation with China on this day. There would be no lifelong nightmares of having shot down one of your own.

The F-105 wingman? He had done all he could do. He had done his duty.

The nightmares would be his.

Almost six

Naha Air Base, Okinawa, 15 June 1965

We had lost two of our squadron pilots while in Okinawa. The day we left, my designated flight leader seemed determined to make it six.

The weather was really lousy. The ceiling was low and the visibility was probably about a half mile outside the clouds. We constituted a squadron of F-4s departing in four flights of four within minutes of each other. We were to join in a flight of sixteen for our flight to Honolulu, a distance of about four thousand miles. Less than a mile after takeoff, my flight leader for the day severely tested all my flying skills and reaction times.

I was in close formation position on my element leader's left wing. We had taken off in two-ship formation, and I was tucked in tight on his wing because of the bad weather. One of the most dangerous things that can occur in formation is to become separated from your flight leader in the clouds. We had just raised our gear and flaps upon visual signals from the lead aircraft. I sensed it was almost time to deselect our afterburners because we were obviously going very fast already. I was having to trim the nose down rapidly, and the aircraft was getting increasingly sensitive on the flight controls. Just as I asked my backseater our airspeed, the pilot in the lead aircraft jerked his engines out of afterburner with no signal and simultaneously rolled rapidly toward me.

Suddenly, I had fourteen thousand pounds more thrust than he did and a face full of his F-4 at the same time. I jerked the throttles out of burner toward idle, opened the speed brakes, shoved left rudder, and left forward stick. My hands and feet were all over the place trying to miss this idiot. Then, just as I got my Phantom semistabilized in our turn, he evidently realized what he had done and snapped his wings back level. At that moment, as I was pulling back up on his wing—so I wouldn't lose sight of him in the weather—he rolled swiftly into me again!

I guess once wasn't good enough for him. It was as if he wanted to see if I could miss him again. I did, but I couldn't believe that I was having to.

I don't have a clue to this day as to what his problem was. Maybe he didn't get any sleep the night before. In any case, his brain was somewhere other than his cockpit. He was supposed to have continued straight ahead for several miles before turning. He didn't. He was supposed to have signaled me before he pulled the engines out of afterburner. He didn't. He was supposed to have come out of afterburner at two hundred and fifty knots. He didn't. It was well over three hundred knots according to my backseater. A good flight leader rolls smoothly into and out of turns, especially in those weather conditions. He didn't.

That guy was supposed to be a highly qualified flight leader. He wasn't. He was supposedly a fighter pilot. He wasn't.

He was a minimal pilot who, unfortunately, happened to be in a fighter. Sometimes they just slip through the cracks. Thank God he didn't go to Vietnam with us five months later. Thank God we soon got on top of the weather, where I stayed away from him for the next three thousand nine hundred and ninety-eight miles. Thank God that there weren't six pilots from our squadron killed in Okinawa.

It sure wasn't my flight leader's fault!

He did *his* damndest!

Go west, young man, about ten thousand miles

MacDill AFB, Florida, 14 November 1965

Jeff was so solemn. He was dressed up in his own miniature flight suit with all the proper squadron and Phantom patches. Even an official name tag with embossed pilot wings was sewn to the chest. His mother was taking our picture together beside my F-4. He was extremely serious, and I just could not get him to smile. He kept looking around at the activities surrounding the Phantoms near us and appeared very concerned. Perhaps at two he was already smarter than I. Here I was, getting ready to leave for a year, and my excitement blanked out all other emotions. Possibly that was normal when you are heading to war, possibly not. But that's the way it was for me. Jeff looked frightened and confused as I climbed into my cockpit. His mother just smiled and waved good-bye.

Slowly I moved the throttles forward, then outboard as Gene (my four-ship element leader) signaled for the second time. The afterburners lit, and I felt the Phantom surge forward. I constantly adjusted the throttles to match the speed of the wingtip a few feet to my left. We were airborne. Our next stop, Honolulu, was only ten hours, forty minutes, and seven air-to-air refuelings away.

The first hitch came over Mississippi. Gene moved to the left wing of our KC-135 tanker. As I slid into the "precontact" position, fifty feet behind the refueling boom, I flipped the switch behind the throttles to open the air-refueling door on the top of our fuselage. The ready light did not come on.

The boom operator confirmed my suspicions. "Whiskey Three Two, your door is still closed," he called.

"Roger, I'll recycle," I answered. Still no light. "Boom, my door is stuck. It's probably cold grease. How about tapping it a little with your boom?" I asked.

"Sorry sir," he replied. "SAC regulations prohibit it."

"Come on boom, just one little tap, and it'll pop right up," I said. If he didn't, I knew I'd be landing at England AFB, Louisiana, and waiting for days for a new "fighter drag" to be coordinated. I don't know why, but it seemed important to me not to be late for the war. Like I said, I think Jeff was already smarter.

"Whiskey Three Two, this is the tanker task force commander. Go ahead and move to contact position. Boom, I'll take full responsibility."

"Yes sir. Whiskey Three Two, you're cleared to contact position," said the boom operator. One tap did it. The ready light beside the center windscreen illuminated. The boom operator hooked up immediately, and the tanker co-pilot began the fuel transfer into my tanks. Honolulu here we come, I thought.

The entire western United States was as clear as a bell as the miles dropped behind us at nine per minute. The Sierra Nevada slid underneath, and in the hazy distance, we could see the California coast near Santa Barbara. Our next rendezvous with the tankers was one hundred miles off the coast. Five KC-135s are pretty easy to find on radar, and there they were, little green dots of light, forty miles dead ahead. My backseater was flying the airplane, and I turned my head to the right to check on the progress of the two F-4s on my wing.

"Twenty-four miles, starting left turn," the tanker leader called.

"Roger," our squadron commander answered. His voice sounded tired, I thought. I was feeling a little weary myself, and I was ten years younger. Besides that, he was leading a sixteen-ship flight of Phantoms. That creates a lot more stress than just flying one.

"Whiskey One One, this is Four Three," came the call.

"Go ahead Four Three," the commander answered.

"Sir, my bird won't take fuel," came the reply. It was hitch number two. Our squadron commander made his decision immediately.

"Four Three, head for Point Mugu Naval Air Station. It's the nearest divert base. Four Four, you go with him. Call me in Hickam tonight. Good luck."

"Roger One One, Four Three's splitting. See you quick as we can." Four Three didn't sound all that unhappy about his delay, I thought. Our commander's quick decision impressed me. He knew that every minute took Four Three and Four Four eight miles farther from land at our present speed, and low fuel can happen very quickly. In an F-4, we were never very far from a low-fuel state.

Two-and-one-half hours later we had just finished air refueling number seven. Honolulu was only two hours away. Beautiful, sunny, balmy Hawaii. As we accelerated back to our normal cruise speed, the tanker commander called, "Whiskey lead, Tank Zero One."

"Go ahead, Zero One," our leader answered.

"Got some bad news, I'm afraid. Hickam weather is six hundred overcast and one mile in rain," he said. I couldn't believe it! Seven hours and fifty-five minutes after takeoff, two hours to go, and we were going to have to accomplish an instrument approach with fourteen F-4s. I suddenly felt ninety years old. I hadn't realized that I was so tired. I told my backseater to fly the airplane while I took a "go pill." I reached down in my flightsuit pocket and found the pills the flight surgeon had issued all of us. I didn't even like taking aspirin, but I had to do something to wake up. My eyes were so tired that I was having trouble focusing them. I put two "go pills" in my mouth and washed them down with some chocolate milk that I had left over from my flight lunch. About ten minutes later the pill had me wide awake, but with an unusually edgy feeling.

"One One's reducing to eighty five percent," our flight leader called. I could see the top of the overcast below us. Slowly, Gene rocked his wings, signaling for close formation. As I moved in to normal wingtip position, our squadron commander called again. "Two One, Three One, and Four One, take spacing. We'll go in with three four-ships and a two ship. The weather is still six hundred feet and one-mile visibility in rain. Keep it in tight guys. We don't want any lost wingmen. We don't have extra fuel to play with. Also, our tankers are right behind us, and they're hurting for fuel also. This weather was totally unexpected."

"I'll say," I thought. Our forecast had been for clear skies and sixty miles visibility. Hitch number three.

At least the turbulence during the descent was minimal. As we dropped below the cloud base on Gene's wing, I saw the Pearl Harbor Channel pass beneath us. Immediately, Two Three and Two Four began dropping back to permit them to land behind us. Gene kept our speed up to help them achieve adequate spacing. At one mile, I began to slide slightly forward, and I knew Gene was reducing power for landing. Out of the corner of my eye, I could see the runway rising to meet us. Gene raised his F-4's nose slightly, and I matched it.

A roostertail of water enveloped my leader's F-4 as it touched down. I grabbed my dragchute handle and rotated it aft. I felt the tug on my shoulders as our opening chute began separating us from Gene's Phantom. His backseater was watching for our dragchute. As soon as he saw it, their chute immediately deployed and blossomed.

Our antiskid cycled slightly as we hydroplaned and then began slowing. Gene cleared the runway in front of us, and I saw the four F-4s of the first flight moving down the taxiway. As I hurriedly cleared the runway, I looked back and saw Two Three and Two Four halfway down it. Water was still kicking up behind them like two racing boats. Out on final approach, I could see two more landing lights just breaking out of the overcast. Eight Whiskeys were down with six more to go.

As I slowly swung my left leg over the canopy rail, I had the strangest sensation. The "go pill" had my mind and upper body wide awake, but my legs and butt were stiff and almost asleep. It was weird.

Day one. Three hundred and sixty four to go.

"Boy this is fun," I remember thinking sarcastically.

Where the sun rises first on America

Hickam AFB, Hawaii, 15 November 1965

"Doc, you've got to be kidding. I'm so tired that I can't stay awake and you're making me take a sleeping pill," I protested. (They were called "no-go" pills.)

"Cook, I don't want you lying here at 2 A.M. wide awake," he said. "It's over seven hours to Guam tomorrow. Besides that, some of you wild men will be downtown if I don't knock you out, so take it. Doctor's orders." I took it. Ten minutes later, the pill took effect. I bent over to take off my boots and almost hit the floor head first.

The next thing I remember, it was the "Doc" again.

"Up and at 'em Tigers. Steak and eggs in thirty minutes," he announced.

"No wonder druggies don't know what day it is. I swear that night only lasted ten minutes," I thought to myself.

My backseater and I sat down in our assigned seats for the preflight briefing on the second leg of our deployment. Our destination was Anderson AFB, Guam, a Trust Territory of the United States. Because of its location west of the international date line, it is "Where the sun rises first on America."

One of our other aircraft commanders turned around and said, "Did you guys hear about the tankers yesterday?" I shook my head no, and he continued. "They were really hurting for fuel after they gave us the extra that they did. They climbed to forty-five thousand feet and long-range cruised. It's a good thing they did. When they finally got down, one of them only had twenty minutes of fuel left."

As I thought about what he had just said, I remembered some of the things I had heard coming out of Southeast Asia about some of the KC-135 pilots. Several had been credited with fighter saves involving some gutsy flying over places where they weren't even supposed to be. Here we had only been out one day and already had experienced some of their obvious dedication. I was glad to know that there were balls in other cockpits besides fighters. That wasn't an easy thing for a fighter pilot to admit, though.

"Let me have your attention, guys." Gradually the noise subsided and the colonel who was the task force commander, walked out. We all came to attention, and he told us to sit down.

"Well troops, yesterday went pretty well. We got fourteen out of sixteen here. The broken jet is fixed in Point Mugu and those bandits will be over in a few days if we can get them out of jail." I never did know if the colonel was kidding, but knowing them and their wingman, there was a good chance he wasn't. Anyway, he got a good laugh. He continued, "I've been in touch with Guam this morning and everything's ready for you. The weather is good and expected to remain that way. Of course, so was Honolulu yesterday. By the way, who made the touch and go landing at Barber's Point Naval Air Station?" (Barber's Point was just a few miles short of the runway at Hickam.) No one spoke. "No one's confessing, huh? Well, the Navy is pissed and said if you did it again, they'd send you to Vietnam!" There was more laughter as the commander walked off the stage. I never did find out if that one was true either.

"Which number are you, sir?" the crew chief asked as he helped me strap in.

"I'll be number ten to taxi," I answered as I plugged in my G-suit hose.

"Sir, why are you sitting on your G suit instead of wearing it?" the young sergeant asked.

"Well, I'm not planning on pulling many Gs, but when I push the G suit test button, it massages my butt and upper legs and helps keep them from going to sleep," I explained. I didn't know whether he believed me or not, from the way he looked at me, but it was true.

"See you in a couple of days at Cam Ranh Bay, sir," the crew chief said and waved as he unplugged his interphone cord. I checked my lineup card to confirm that the Phantom taxiing in front of me was Gene's. He looked at me as he went by and nodded. I pushed the throttles up, and my F-4 moved forward very sluggishly. The hottest fighter in the world wasn't so hot when it was loaded to the gills with full internal fuel and three fully laden external fuel tanks.

The departure from Honolulu was going to be a buddy formation with our tankers for the first and second refueling. I recalled the serious looks on the faces of the tanker pilots as they had briefed their portion. They were getting ready to accomplish probably the most dangerous part of their job, getting a fully loaded KC-135 off the ground at Hickam and beginning an immediate right turn to avoid Waikiki and the hills beyond. If nothing went wrong with an engine or their water injection system, which added extra thrust, they could make it. If something went wrong, they probably couldn't—simple math.

Nothing had gone wrong so far. They were airborne, in a manner of speaking, and beginning that dangerous right turn just above the water. The pilots had to finesse all the speed that they could out of their tankers before the water in the injection system ran out one hundred and ten seconds into the takeoff. By that time, the flaps had to be retracted, and the speed had to be adequate for the climb, or no climb. (All these critical procedures and dangers could have been avoided for over twenty-five years, except for a certain cigar-chewing four-star general who was too impatient to wait six months for delivery of a KC-135 with engines that totaled thirty two thousand pounds more thrust. Incredible!)

Gene's head dropped forward, and I released the brakes: day two. The afterburners lit, and I felt the now-familiar thrill of brute power begin again. We were airborne, gear up, flaps up, and afterburners out. Just the opposite of the KC-135 tankers, our speed problem was going too fast, too soon. It wasn't likely, heavily loaded as we were that day, but the engineers had installed an electrical switch to automatically retract the flaps at two hundred and fifty knots if we didn't.

As we passed three hundred knots, Gene spread us into route formation and signaled for a power reduction. I moved into a wider formation position and looked ahead. There he was. We had already caught our tanker as he was struggling through two hundred and fifty knots. I felt sorry for them and remembered again the cigar-chewing general. We were passing five thousand feet and our tanker was finally reaching its climb speed of two hundred and eighty-five knots.

We were closing slowly on the lead tanker and his fighters. They were climbing ten knots slower than us, so it would be a while because they were about five miles in front of us.

"Whiskey 41, all chicks in the green," came the call from the last flight. Fourteen F-4s and five KC-135 tankers were all airborne and all operating normally. So far, so good. Only a little over three thousand miles of water to cross until Guam, except for Wake Island, that is.

It amazed me that the Pan American Clippers ever found it in the old days. Wake Island is located about two thousand miles west of the Hawaiian islands. It looked incredibly small from thirty-five thousand feet. It's essentially a runway on one reef connected by a small taxiway to the tiny ramp on another reef.

Only thirteen hundred miles to go: We had been flying about four hours since we left Hawaii, but had flown right into the next day as we crossed the international date line.

Strange. I have never been able to fully grasp that concept.

Bombers, bombers, everywhere: Thank goodness, I never had to fly one of the damn things!

Anderson AFB, Guam, 17 November 1965

I had never seen so many B-52s in one place. They were sitting droop-winged all over various parts of the huge ramp. Others were taxiing to or from the runway. Still others were entering the traffic pattern for landing. Along with the "Buffs," there were KC-135s, C-130s, C-124s, and various civilian contract DC-8s and Boeing 707s. I figured the cement contractor for this place was rich, happy, and probably retired to his yacht in the Caribbean.

We parked our F-4s and deplaned. Today had almost seemed like a walk in the park in comparison to yesterday, or was it day before yesterday because we had crossed the date line? Anyway, three hours less flying time and excellent weather had made a huge difference. We filled out our aircraft forms and climbed on the bus to go to our billeting. A hot shower and some hot food sounded like heaven to me right then. Only slightly over two thousand miles to go tomorrow and we'd be—home? I admit that the nearer we got to South Vietnam the more serious my mood became. Anyone who ever tells you that they weren't at least a little nervous about going into combat is one of two things: either a liar or certifiably nuts. In either case, it's an excellent idea to avoid that sort of person.

I don't remember much about the short stay at Guam. I do remember that a lot of time was spent briefing us on the physical characteristics of our destination airbase, Cam Ranh Bay, South Vietnam. I guess the cement contractor hadn't found it yet. The runway was an aluminum mat laid out on the sand, ten thousand feet long and one hundred and two feet wide. All the taxiways and the ramp were aluminum also.

We arrived over the peninsula containing Cam Ranh Bay and made a wide sweeping descent around it. From the air, it looked like a tropical paradise with the white sand and blue water. We entered the initial pattern in four-ship formations and made overhead fighter breaks for our first aluminum landing. It really wasn't so bad. A little bumpy and narrow, but not bad. Little did we know that some of our most exciting moments in Vietnam were going to be involving this long piece of metal matting, but that's another day. This was day three, or was it four because of the international date line?

As we shut down the engines, some of the extra pilots who had come ahead on cargo aircraft met us and tossed us cold beers. They even had some cold Cokes for teetotalers like me. They didn't need many Cokes! Anyway, it was the welcome to our "all-expenses-paid year's vacation in exotic Southeast Asia."

Three-hundred-and-sixty-two days to go, or was it three-hundred-and-sixty-one?

Welcome to the war

Cam Ranh Bay Air Base, RVN, November 1965

There was sand everywhere. When you walked in it, it squeaked. When you took a step forward, your other foot slid, no, *rolled* back a half step. The grains of sand were almost perfectly round, like little ball bearings. We were now living on a sand peninsula, but get this—the engineers had to import sand! It seems that the local stuff had been ground so round by centuries of the South China Sea's hydraulic action that it would not make concrete. The cement would not stick to it like normal sand, which has uneven rough edges.

It was a major effort to walk anywhere. I swear that you burned more calories walking to the cook tent to get chow than you consumed in your meal. It was also very hot and humid, and we were usually dressed in our flight suits, guns, and boots, which were not cool in the temperature sense of the word. I weighed two hundred and seven pounds when I got to Vietnam. I weighed one hundred and sixty-seven when I left a year later. I don't highly recommend it as a way to lose weight, but it was obviously effective.

One thing that the sand was good for was filling sand bags. We spent lots of our early leisure time digging foxholes and filling sand bags. Our base was somewhat easy to secure because of its topography. We never had to use our foxholes, but we slept better knowing that they were right outside our barracks and covered by the sandbags. I'm the first to admit that our situation could have been much worse in other locations. Another factor in our favor was that we were guarded by a contingent of ROK Marines. That's "Republic of Korea" for the uninitiated. These guys were so fierce and had such a "bad" reputation that the Viet Cong reportedly were scared to death of them. For whatever reason, we were never attacked by the bad guys while I was there. An added bonus from the ROKs was that they were great "Lock and Loll" musicians and provided us with a lot of entertainment, for a slight charge, of course. "Hang on Sloopy," whatever that meant, was one of their best and my personal favorite.

Our Fighter Wing consisted of four squadrons of F-4Cs. Our work schedule the first few weeks was three days of flying and one day of building. We were continually building. We built a dining hall. We built an officer's club. We built wooden sidewalks to connect them all. We built rooms inside of our barracks with four pilots to a room. Each room had a two-ton Chrysler air conditioner. (Our electrical power supply came from a huge generator originally designed for a Strategic Air Command missile silo.) We could turn those babies on and have icicles hanging from the ceiling in ten minutes. I don't know how we kept from getting pneumonia with such huge differences in outside versus room temperatures, but it sure made for good sleeping after a day in the cockpit. I must confess that I sometimes felt guilty knowing that the U.S. Army guys were out there in the boondocks, not only with the enemy, but with mosquitoes, snakes, the oppressive heat, and God knew what else. I was very, very fortunate, and no one knows it better than I.

I don't remember very much about my first few missions. I do remember that I was very unsure of what to expect and that the unknown is usually what we worry about the most. I do remember walking out to the airplane on my first combat mission, though. My formation flight lead was one of the four flight commanders in our squadron. I remember his name, but won't reveal it. We were laughing and talking a little nervously about our upcoming baptism of fire. I looked down and noticed that he was missing a vital part of his flight gear as we crossed the hot aluminum ramp. I asked him if he felt any nerves, and he assured me that he did not. (Any fighter pilot would have said the same thing, including me.) So I asked him where his helmet was, and I can still remember the sheepish grin on his face after he looked down at his empty left hand. He just shrugged and started back across the hot ramp to the equipment room.

Which ones are the bad guys?

Tan Son Nhut Air Base, RVN, December 1965

Seventh Air Force was our boss in Vietnam. Lieutenant General Moore was the Commanding General of the Seventh. Its headquarters at Tan Son Nhut Air Base at Saigon contained a large planning section for air strikes "Out of Country." It was where the latest intelligence concerning North Vietnamese defenses was received and processed. There was a large plotting board displaying all known surface-to-air missile sites and antiaircraft gun positions. These plottings were in a constant state of change as updated intelligence information arrived and was processed.

I returned from an airstrike in early December and was met by my operations officer. He had a grin on his face and said "How about a little vacation, Jerry?"

I knew better than that, so I said, "Yeah right, major. What's the catch?" He handed me the piece of paper. I had been assigned temporary duty at Tan Son Nhut for ten days. It seems that because of the raids being directed from Seventh Air Force, there was a requirement for an "expert" from each assigned aircraft type to help with the planning.

I said "Why me?"

He just laughed and said, "You're the most junior expert." I had to be there the next day. It was short notice, but that was usually the way things happened in Vietnam.

After a noisy C-130 ride to Tan Son Nhut, I reported as ordered to the intelligence division. They gave me and the other expert reporting that day a short briefing, told us where we were to be billeted, and we went to work. Our job was to study the next day's target assignments for all fighter missions into North Vietnam with regard to forecast weather and enemy defenses. We would choose possible routes from and returning to the supporting fighter bases. Depending on the target, we selected the type of bombs and the estimated number required, number of fighter bombers required, number of MiG cover fighters required, fuel required, and air refueling tankers required. It was an involved undertaking and one of extreme importance because any mistakes we made could be reflected in the loss of some of our fellow fighter pilots. The responsibility weighed heavily, and I was glad that it was an assignment of short duration.

The other pilot reporting that day, and my roommate for the ten days, was an F-105 pilot. His first name was also Jerry. He had been in Southeast Asia for several months, having flown most of his missions into North Vietnam. At the time, the F-105s were flying seventy-five percent of the bombing missions there. He had seen a lot of intense action, and several of his squadron pilots had been shot down, including one of his wingmen. At the present rate, there was a one-in-four chance you would be shot down before completing one hundred missions over North Vietnam.

Jerry was pretty quiet about it, and I didn't ask many questions, although I wanted to know more. He had developed a somewhat fatalistic attitude and didn't smile or laugh very much. In fact, he seemed to be in deep thought most of the time. One day I jerked him back onto a curb in downtown Saigon just before he would have been hit by a taxi. When I told him to be careful, he just looked at me and shrugged. I'm not sure he thought that I had done him a big favor by saving him for more combat flights.

Part of our inbriefing had focused on some of the current Viet Cong tactics. We were told to avoid crowds at all costs and to not let anyone of any age brush up against us. It seems that two current tactics were for some kid to come up behind you and ram a fishhook between your shoulder blades. Attached to it was a short fishing line with a hand grenade on the other end. After placing it where you couldn't reach it, he would pull the pin and run. Another was to brush up against you and inject you with cobra venom. I don't remember how many steps they said you had left, but it wasn't many. I can't recall seeing any reports of incidents like these in our newspapers, but I know firsthand that there were a lot left unreported, for whatever reasons.

Another part of the briefing concerned the bounties on our heads. It seems that the Viet Cong were offering money for the capture of American pilots. I believe the amount on a captain like myself was ten thousand dollars. If they captured you, the report was, they would skin you alive, then cut off your head and parade it through the villages on a stake for propaganda purposes. I remember seeing a picture of one such incident.

Saigon was a very crowded city, and we were staying down by the river at an old French-built hotel named the Majestic. In order to get to the airbase and back every day, we had to utilize taxis. They were everywhere, and most were little Morris Minor sedans painted in various bright colors. We always tried to finish our work early enough to beat the crowd of workers vying for taxis outside the gate at Tan Son Nhut Airbase at quitting time. This particular day, we were later than usual. When we got to the taxi pick-up point, there were people everywhere. We got in one of the lines and began waiting for our turn.

Suddenly, I heard, "Captain, captain, over here." I looked in the direction of the voice and saw a black Mercedes four-door sedan. The driver was a young Vietnamese in a khaki uniform with no visible markings. He then said, "Come, I give you ride to hotel. I charge you same as taxis, one hundred and fifteen piasters."

I said "You don't look much to me like a taxi."

He replied, "Oh I not. I borrow car from friend to make extra money."

I looked at the F-105 pilot. He said, "Why not? It's going to be an hour otherwise." I wondered if he was thinking the same thing that I was. I got in the left rear seat anyway. As we drove away from the crowded taxi stand, I remember the driver looking in his rear view mirror and smiling. We had been in Saigon long enough for me to recognize the general direction to downtown. The Mercedes driver was not heading there. He was driving in a northerly direction.

I leaned over the back of the front seat and said, "Take the next right turn and head toward downtown."

He gave me a big smile and replied, "Oh no captain, much better we go this way."

I looked out the windshield of the sedan and repeated, "Take the next right turn and drive us straight to the Majestic Hotel."

He passed by the next corner and started to say something else. I stuck the barrel of my .38-caliber "Combat Masterpiece" in his right ear and said, "Turn around now, and take us to town."

He opened his mouth to say more. I cocked the pistol with the barrel still sticking in his ear. He turned around and drove us, very carefully, straight to our hotel. When we pulled up at the front, I uncocked the pistol and then removed it from his ear.

The other pilot got out and walked up to two military policemen standing on the sidewalk. I placed the gun back under my folded newspaper.

The look on his face was incredulous as I handed him one hundred and fifteen piasters and said "Thanks for the ride." (Hey, a deal is a deal!)

Of course, by now a military policeman was opening his driver's door for him.

Sidewinder 21

Tan Son Nhut Air Base, RVN, December 1965

I was to head back to Cam Ranh Bay Air Base in a day or so. I had felt much more at ease there flying combat than with the crowds of Vietnamese in Saigon with unknown political leanings and the "taxis." Just prior to leaving Saigon, I ran into an old friend that I knew from my instructor days in the Air Training Command. I had just ordered a hamburger at the Tan Son Nhut Officer's Club, when I heard my name being called. I looked over and there he sat, alone at a table. I picked up my food and walked over and sat down after shaking hands.

We caught up briefly on each other's recent history, and he asked me what I was doing there. I told him, and he asked me how it was flying with people shooting at you. I told him most of the time, so far, I didn't even know that I was being shot at. I also related that the first time I did see ground fire, it made me angry. I laughed and told him that it was probably an illogical reaction since I was shooting at them too, but nevertheless, that was it. He asked me if I was scared to fly combat missions. I said that on the first few I had been apprehensive, but now I had pretty much adjusted and just remained alert.

He said, "No, I mean are you afraid that you're going to get killed?"

I looked at him and replied, "No, I don't think that they can kill me. I believe that I'm going to make it through my tour just fine."

He looked at me sadly and said, "Jerry, I think I'm going to get killed." I could tell he was serious and tried to talk to him about having self-confidence and positive thinking. I told him that I thought having confidence that you would make it was a very large part of the equation.

He shook his head and said, "I've tried to shake this feeling, but I've had it ever since I got this assignment. I even canceled all my regular life insurance and took out four hundred thousand dollars worth of term life." I tried talking some more to him, but I could tell that I wasn't doing any good. I changed the subject. He was going to be a forward air controller (FAC) in an area where I flew airstrikes occasionally. He was going to be one of the FACs with a Sidewinder call sign. I had to get back to the intelligence section, so I shook hands again and wished him good luck and told him to try to quit his negative thinking. I told him that I'd be listening for him when I flew into the Sidewinders' territory.

Probably a month later I was on a strike mission and was assigned a Sidewinder FAC. It sounded a little like him over the radio, so I called, "Dave, is that you?"

He immediately answered "Yes," and we carried on a short conversation between my bomb passes. As we left the area, I said I'd talk to him later and changed radio channels. Sure enough, about a week later I had Dave again for my forward air controller. We shared some small talk and he said he was doing fine. I said, "So long" and signed off.

Several weeks went by before I returned to the Sidewinder FAC area. It was Dave's callsign and I said "Hi Dave," just before my first pass. He didn't answer, and I assumed that he hadn't heard me. Before rolling in for my second pass, I said, "Hey Dave, it's Jerry."

The FAC called me and said, "Are you calling for Dave," and used my friend's last name. My stomach knotted. I said that I was.

"I'm really sorry to tell you this, but Dave was shot down and killed last week."

Seagulls and paper tigers

Cam Ranh Bay Air Base, RVN, 1966

Every flying unit has them. Without a shooting war, they may never be found out. Combat usually exposes them. I mentioned in an earlier story about someone who was not a fighter pilot, but was a pilot flying a fighter plane. I think a true fighter pilot is *not* fearless. I said before to avoid anyone who claims to be unafraid because that person is either a liar or crazy. My friend in the previous story was not flying a fighter plane, but he had a fighter pilot spirit. He was fearful that he was going to get killed, but did his job and flew combat anyway; however, there is no disgrace in being so scared that you can't fly combat. The disgrace is in not admitting it and trying to cover it up with lies.

One of our pilots was completely honest about his fear of being shot at. He was an excellent pilot, but simply could not control his phobia. The F-4, one of the more difficult aircraft in the inventory to fly, did not frighten him. He simply could not overcome his terror of combat flying. He went to our flight surgeon after several missions and explained his plight. The flight surgeon did not ground him as he could have. We needed test pilots to fly our battle-damage-repaired F-4s and to deem them airworthy for the rest of us. That was at times a dangerous job, but he did it and did it well. He was up front and honest about his problem, and was never shunned or thought less of by any of us. In fact, we respected him for his own kind of courage.

There was another pilot who was not so honest. He looked as tough as nails. He stood about six-feet-two inches and weighed around two hundred and twenty pounds. Stateside, he was brave and tough. His paperwork looked good. He was a fighter pilot as far as any of us could see, especially to hear him tell it.

There seemed to be something wrong with a lot of the F-4s that he was assigned to fly. Several four-ship formations took off as three-ship formations instead. I'll wager that if the records could be checked, his abort rate went up in proportion to the expected difficulty of the mission. He barely had twenty-five missions when the rest of us had around twice that. We started referring to him as the squadron "seagull." That was a reference to a pilot who just "shit and squawked and had to have rocks thrown at him to get him to fly."

One day he was gone. It turns out that he had "developed" an "exotic blood disease" and was at Clark Air Base in the Philippines for medical evaluation. It seems that he had a severe rash on his upper legs and it had started bleeding. (One of our other pilots testified that he had walked in on him one day, and he was rubbing his rash with a wire brush, which he quickly hid from sight.) If there was any pilot at Cam Ranh Bay without a rash, I'd be extremely surprised. It was the predictable result of the heat, humidity, and tight parachute straps. I suspect that the seagull's rash was no different than anyone else's.

One of the pilots from another squadron returned from Clark Air Base where he had been test flying battle-damage-repaired F-4s. He had been at the Officer's Club Bar and one of the American school teachers based at Clark spotted his F-4 pin in his lapel. She said, "Oh, do you fly F-4s?"

He proudly said "Why yes ma'am, I do."

"Then you probably know" (she called our seagull's name).

"Yeah, I know him," he said.

She then proceeded to tell him about how wonderful and brave the seagull was when he won the Silver Star. That really got the other pilot's attention, and he began to ask her questions about what the seagull had told her.

It seems that he was leading a flight of F-4s against a heavily defended target in North Vietnam. Inbound, the seagull's plane was hit by antiaircraft fire, and he was wounded in the leg. But he had undauntedly continued to his target and destroyed it. Because of that, he had received the Silver Star.

The astonished pilot said, "So, he was hit in the leg, huh?"

She assured him, "Yes. In fact, that's why he was at Clark. He was getting his wounded leg attended to." She also confirmed that he was limping bravely all over the base.

Shortly after we received this story of "heroism" back at Cam Ranh Bay, we heard further news. Word was he had received a disability settlement and medical discharge from the United States Air Force.

Shot down, by a nurse

Cam Ranh Bay Air Base, RVN, 1966

When we first arrived at Cam Ranh in November 1965, things were pretty basic. We were on a peninsula consisting almost entirely of sand. I don't ever remember getting a glimpse of solid ground, except when we were digging our bunkers. It was a long way down to the ground beneath the sand.

Our meals were served out of a mess tent, and there was another tent for eating that was equipped with tables and chairs. The food was delicious. I really do think it was that good, but another possibility was the sand. I mentioned in another story that the sand grains were round. When you walked in it, as one foot went forward, the other slid backward several inches. It was like walking on ball bearings. Consequently, by the time you walked any distance in the stuff, you were worn out. I wouldn't be surprised if we burned more calories walking to the mess tent than we consumed when we got there. Perhaps that is why it tasted so good. One thing was sure. There weren't many fat boys around, at least not for long.

It didn't take long for us to find a solution. We decided to build a combination Officer's Club and mess hall. We connected it to the barracks areas with board sidewalks. It was great. It not only was easier to walk to chow, we weren't soaking wet with sweat by the time we got there. Of course we stopped losing weight at such a fast rate, and somehow the food didn't taste nearly as good as it had in the old mess tent. But those boardwalks were nice.

Cam Rahn Bay had approximately five thousand men when we arrived and no women that I was aware of. Then the base hospital arrived—with the nurses. I think the ratio was about one thousand-to-one "males to female." Shortly thereafter, the ratio changed a little when the Red Cross "girls" arrived. It then became about five hundred-to-one.

We had a lot of bachelors and a few guys who weren't married, but whose wives were. The race was on!

One evening I had finished dinner and was talking shop to another pilot in the bar area. Near us, sitting at the bar was one of the wing fighter pilots who was known to us but will remain unknown to you. He was deeply engaged in conversation with one of the nurses. It seems that he wanted her to go for a "walk" with him on the beach. He had obviously been trying to influence her decision by buying her a few drinks, which she had obviously accepted. She was ready. The problem was a third person sitting immediately on the other side of the now willing nurse. It was another nurse. She was a tiny blonde who couldn't have weighed ninety pounds. She was about five feet tall. She was trying to get her tipsy nurse friend to go back to her quarters and get some sleep. The friend didn't want to go. She had been convinced that she needed to go to the beach with the fighter pilot.

The pilot stood up, and his selected "walking" companion stood up with him and took his arm as the little blonde continued to protest. The happy couple started for the front door of the club, but the blonde nurse followed. The pilot was getting very perturbed by now and told her to go mind her own business. She said something unintelligible back to him and tried to grab his companion by the arm. She pulled away, and they all went outside. I thought this was getting pretty interesting, and, since we were always starved for entertainment, several of us followed the trio outside.

They were oblivious of us. They were standing on the boardwalk about ten feet from the door. The pilot had "his" nurse by one arm, and the little blonde nurse had her by the other arm. They were arguing and pulling the nurse in the middle back and forth, back and forth. I think she started singing something about that time which indicates the condition she was in. This was getting good!

The fighter pilot finally got mad and made his "fatal" mistake. He called the little bitty blonde nurse a *very* bad name. She let go of her still singing friend's arm and knocked the five-foot-eleven-inch, one-hundred-and-seventy-five-pound macho fighter pilot right off the boardwalk and into the sand. She then took her friend's arm, said "Let's go!" and departed the "combat zone." We all cheered, that is except for the "crashed and burned" fighter pilot sitting in the sand holding his nose.

The next day word got around the base that the pilot's nose had been broken and he had gone on a "Rest and Recuperation leave," better known as an "R and R," especially since he couldn't fly for a while with a "busted beak" anyway. I think it probably had more to do with a "busted ego."

If I recall correctly that was the only injury he sustained in a year of combat.

"War is Hell!"

Number twenty-five

Cam Ranh Bay Air Base, RVN, 1966

Whether you saw enemy ground fire depended on a lot of factors. Weather, ground cover, smoke, dust, and size of weapons being fired were some of them. Sometimes we saw lots of ground fire, sometimes none. Many times our forward air controller, if we had one, would report ground fire when we couldn't see it.

Mission number twenty-five for me was a strike on a large number of Viet Cong holed up in trenches near the South China Sea, south of Cam Ranh Bay. Our forward air controller had stumbled onto them because some of them couldn't resist taking shots at him as he flew past on patrol. We were scrambled from runway alert and proceeded south along the shoreline while he circled wide of the area and took stock of their location and probable numbers. Shooting at him proved to be the dumbest thing they could have done. If they had not, he might have droned on past and never have seen them. It also proved to be one of the last things they ever did.

As we arrived in the area, the forward air controller briefed us on the situation. There were an estimated several hundred Cong in trenches near the beach, which he would mark for us with a smoke rocket. He received a lot more ground fire as he surveyed the area. Evidently they had realized their mistake and decided that they needed to shoot him down before he was able to call in his fighters. They failed.

He rolled in suddenly and fired a smoke rocket toward the trenches. They immediately lit up like Christmas-tree lights as the enemy started firing with everything they had. It looked like mostly thirty-caliber stuff. The ground fire actually made it easier for us to find the target. Moving at almost five hundred knots, it's sometimes hard to pick out things on the ground. The muzzle flashes made it easy. All we had to do was aim for them. Of course, that also meant that their muzzles were aimed toward you. Our two-ship flight carried napalm, which was perfect for this situation. As I made my bombing runs, I aimed for the heaviest concentrations of flashes. We made four passes in all, dropping on each one. As we left the area a few minutes later, the forward air controller flew over it and reported no ground fire or any other enemy activity. He estimated that we had ninety-five KIAs (killed in action) and the rest wounded. I hadn't taken any hits at all that I knew of.

I parked my F-4 at the south edge of the ramp abreast the white-bladder refueling tanks. I got out and began filling out my aircraft maintenance forms as the ground crew started servicing the Phantom. I felt something strange on my leg and looked down. Jet fuel was squirting out of a hole in the left fuel drop tank that I was using for a desk. My leg was getting soaked. I yelled for the ground crew to stop the refueling and stuck my finger in the bullet hole to plug it. A ground crewman relieved me with a wooden plug, and we began looking for more holes. We found several more and three deep nicks in the forward windscreen made by some kind of flying objects. It would take a few days of patching before this bird flew again.

I spent the next few missions sitting low in the saddle, as if that would have done any good. Oh well, it made me feel better.

Surf's up

Cam Ranh Bay Air Base, RVN, 1966

One of our backseaters was from Hawaii. He loved the water and the beach, and the sand at Cam Ranh's beach was some of the whitest I had ever seen.

I don't like the beach or the ocean. I don't like sand in my shorts, and I don't like the big fish that swim in the ocean that like to *eat* you.

I think that it stems back to a trip I took with my parents to Galveston, Texas, when I was small. We were walking along the beach and saw a group of people in a circle around something. As we got closer, we heard someone talking about the object that everyone was surrounding. Whether or not I heard it correctly or whether my child's imagination "took the ball and ran with it," I don't know.

I thought I heard that it was a shark that had washed up in the small surf with a boy's foot in it. It didn't matter from then whether it was true or not. It was implanted in my head and has been there ever since; therefore, I don't like the beach or the water.

Danny, the Hawaiian native, loved it. He also liked to surf and had brought his board. Almost every opportunity he had—when he wasn't flying—he went to a small cove just southeast of the air base and would "work" the small surf that was sometimes rolling in. It wasn't much of a surf, and it took an expert to ride it. He was good and about the only one around who could do it.

As the story goes, Danny had been surfing for some time in the cove by himself. He had gotten a little tired and just laid down on his board on his back to doze in the sun and float around in the cove. He went to sleep. He was many yards from shore.

The loud roar and whirring noise woke him. It was a U.S. Army helicopter gunship, and it had just buzzed him. He assumed that they were just fooling around and waved at them and shut his eyes. They came right at him the next time and passed over very low. The downwash from the chopper's rotor rocked the surfboard. They came back around in front of him and began a hover. He could see their arms moving, and he thought that they were waving. He waved back. They fired!

The machine-gun rounds whizzed over and behind him and scared the "stuff" out of him. Then he turned to look where the gun rounds were hitting. A huge gray shark was thrashing and rolling just a few feet behind Danny. It was really tearing up the surrounding water.

Here's the description I heard of what happened next: Danny jumped straight up, grabbed his surfboard, and *ran* to shore, although the water depth was well over his head. Of course I don't believe that version, but then, I wasn't there!

The beach to the east of the main base area was big and beautiful. The problem was, the sharks that sometimes hung around just off the sand were also big and beautiful.

Now you would think with the shark sightings from time to time that there would be no problem with keeping people out of the water. You think wrong! Every time that I would wander down to the beach to kill some time, I would see people swimming. There was a shark "watch" organized, and once in a while the swimmers would be called out of the water because of shark sightings.

I don't think I'm wrong when I say there was at least one shark attack and perhaps even a fatality during the eleven months I was there.

I was returning from a mission one day with some unexpended ordnance, which meant that I would not fly a normal overhead pattern; I would fly a "hung-bomb" pattern. It basically was rectangular and stayed offshore and away from anything that the bombs could damage if they inadvertently fell from the jet.

I was just approaching the area of the beach off on my right-front quarter. I looked over and could see lots of people on the sand and several in the water. I looked closer to my position and saw a huge object just below the surface. I remember my first impression was that it looked just like an F-100 fuselage without wings. It must have been at least thirty feet long.

I quickly called the tower and told them to contact the air police on the beach about the great shark stalking the airmen.

Like I said, I *don't* like the beach or the water!

Through the cracks

Cam Ranh Bay Air Base, RVN, 1966

"Now we're going to be operating close to the Cambodian border this morning, so watch yourselves. It's been reported that they have some old Korean-vintage jet fighters. If they try to interfere with us because of our close proximity, remember that we will not be carrying any air-to-air weapons. Our only available tactic against them is to get over them and drop a bomb on them."

When I heard later that he had briefed a flight that he was leading on that "tactic," I thought that he had to have been joking. Those in the flight assured me that he was not.

Everyone in a fighter unit knows who they are. At least the pilots who have to fly with them on a regular basis know. They are the ones who somehow "slipped through the cracks" in the selection and training criterion of the fighter pilot "community." I don't know *exactly* how they do it, but I have a pretty good idea. I do know that there is a failure on someone's part along the line to weed these people out. I think usually there is "influence peddling" somewhere. Political strings are pulled. Calls are made. "Suggestions" are given. Heads are turned the other way and "behold," a minimal pilot ends up in a high-performance fighter plane. It's bad enough to have one of these guys flying on your wing, but with enough time and luck on their part, you end up on theirs.

None of the backseaters who had flown with him wanted to do it again, but they also didn't want to get into trouble. They talked about it among themselves and complained to some aircraft commanders about it, but the subject "fighter pilot" outranked all of them and held a position of power over all of us. I don't think that his immediate supervisors knew the situation was as serious as it was. Actually, I'm not even sure that they even knew that there was a "situation."

Being a frontseater, I was never subjected to riding with him in the same airplane, so I couldn't really appreciate the backseater's plight and concerns either.

However, I did fly in two flights with this guy leading that remain quite vivid in my memory cells. In fairness to him, there might have been other flights that were normal and consequently not necessarily noteworthy in my thirty-year-old memories.

Flight One:

It was really kind of nice being number-four aircraft in a flight of four. You didn't have to navigate. You didn't have to talk much on the radio except to check in. You didn't have to concentrate on being real smooth because no one would be flying on your wing. All you had to do was fly formation off of your element leader ("elephant breeder" as we jokingly referred to it). When you arrived over the target, you dropped your bombs or fired your gun, made sure you didn't run into anyone else, and then flew loose formation home, if the weather was good. It was a "piece of cake."

We were carrying seven-hundred-and-fifty-pound bombs and a Gatling gun. I remember that our target was somewhere between Tuy Hoa and Pleiku. I don't remember what the target was. I do remember who the flight leader of the four-ship of Phantoms was. I also remember that the target was "hotter" (more intensely defended) than we expected it to be.

It was the last pass. Everything had gone pretty much as planned. The other three Phantoms had pulled off the target with number three having just dropped its last bomb. I was slowly raising the gunsight up to the target, and my rearseat pilot was calling off the altitudes as we plunged through them. Our dive angle was about thirty degrees. "Pickle" called my backseater, and I punched the bomb release button on the stick. I "laced in the Gs" to around six and the nose was coming up.

A heavy metallic "thunk" vibrated through our F-4, and the right-hand fire warning light came on simultaneously. As I pulled the right engine back to idle, I called "Four's hit. I have a right engine fire light." I kept the left throttle at full military power and brought the nose of the Phantom around and pointed it at the rest of my flight. They were effecting a loose joinup with the lead F-4 as he headed back toward Cam Ranh Bay.

"Lead, this is three. I have four in sight. I'll drop back and look him over," said my element leader in the third F-4. The flight leader had not uttered a word until now.

"Join up on me three and leave him," came the unbelievable call. I had heard that this guy was a total jerk, but I didn't know he was this bad, until now.

"But lead, I've got him in sight, and I can get to him in a couple of minutes," protested my element leader.

"I said leave him, three. We're low on fuel. We can't afford to wait around," came lead's terse reply.

I looked down at my fuel gauges. It was a damned lie! We had just had a fuel check before our air strike and everyone was pretty even. In fact, as number four, I had been low man, which was not unusual. What was his problem? One thing was readily apparent. He sure wasn't going to let me be his problem. I can remember thinking that he was one sorry S.O.B.!

The fire light was still on with the power in idle, so I shut the engine down. The light stayed on. There was nothing else to do. With no one to confirm whether or not we had a fire, we just had to fly home not knowing for sure.

The flight pulled away and left us. I watched them disappear. On one engine without using the afterburner, all we could manage was seventeen thousand feet and a much slower speed than normal. Finally, a couple of minutes after level-off, the fire light went out. I reached up and toggled the fire-warning test switch, and it worked properly. We felt better.

I was about to get my anger and blood pressure under control when our "flight leader" called.

"Four, say your position and altitude. It looks now like we have plenty of fuel. We'll come back and check you over." What a guy! My blood pressure started to rise again.

I would not answer him. I didn't want him anywhere near me. We landed twenty or thirty minutes after the rest of the flight. I did not go to the debriefing. I did not say anything to the so-called "flight leader" who outranked me. I did not want to go to the "stockade" for what I probably would have done.

In retrospect, I probably should have done it.

He deserved it.

The right engine had to be changed. Shrapnel had shredded some of the compressor stage. There had indeed been a fire.

Flight Two:

For some reason we were just a two-ship flight that day. Normally we were scheduled as four aircraft; however, occasionally someone would abort, and we would end up with fewer birds. The target area was in very rough mountainous terrain. It was to the northwest of Kontum almost right on the Laotian border. A forward air controller was in the area to pinpoint our target for us.

There was a deep valley with a little river running through it. I remember it appeared that there were some white-water rapids. I also recall that the general direction of the river was toward the southeast, I thought.

The target was one of *our* helicopters. It had been shot down by ground fire, and the Army had decided that it could not get in to retrieve the 'copter. The terrain was too rough, and the "natives were too unfriendly" and numerous. The forward air controller briefed that the Army did not want the radio and other equipment in the helicopter to fall into enemy hands. Our job was to destroy the helicopter before that could happen. The weapon of choice was our "pistols" (twenty-millimeter Gatling guns).

The forward air controller did not have to mark the target. We could see the Army chopper on a sand bar in the river where the pilot must have autorotated. He had evidently done an excellent job as it looked relatively intact from my altitude about five thousand feet above it.

The forward air controller got out of the way toward the east and cleared our flight of two in "hot" (cleared to fire). I was turning onto my base leg for my first strafing pass. I watched lead as he began firing and could see the familiar gunsmoke trail from the cannon mounted on the belly of his Phantom. I looked back in the cockpit to recheck a switch position and then rolled in on the helicopter. From my vantage point, it looked like we were aimed down river. As we roared between the mountain ridges and into range, I began to fire. The helicopter was already smoking from lead's gun pass. *"Oh S---!"* I stopped firing immediately and pulled hard on the stick. The ground was rushing up at a tremendous rate.

I roared over the chopper as it was beginning to burn, but I wasn't concerned with the helicopter at the moment. I was too concerned with trying to miss the rapidly rising terrain without stalling my F-4. I learned early in my career that if I found myself in a dire situation where it looked like I might not be able to miss the ground, to *not* try and pull out as high as possible. I pulled out as *low* as possible. That way I have given my jet as much space as exists to keep it from stalling. Once you stall it, you are probably just a passenger on your last ride. (We were also taught that if there isn't enough room left, you "bend over and grab your ankles and kiss your a— goodbye!")

I had screwed up. I had thought that we were strafing "down" the river with the terrain falling away from us. It was a deadly optical illusion, and it had fooled me. I wondered if my flight leader had been affected by the illusion also. If he had, he sure as hell hadn't warned me about it.

We cleared the river by a few feet, and I pulled up to the left for a downwind leg and my second pass. I didn't see lead anywhere ahead of me in the strafe pattern.

"Two, go ahead and strafe by yourself. Lead's off high and wide," came the call.

I looked up and to my right and saw the lead F-4. He was nearly invisible against the morning sun. I wondered for an instant what that was all about, but immediately began planning my second pass. It was going to be a lot different than the first one, I assure you! I had already gotten as close a look at the river as I cared to on the previous run.

"Two, abort the mission and join up. Lead's bingo fuel," lead called just as I was about to turn onto a base leg for my second run. Now I really wondered what was going on as I checked my fuel. I still had plenty for one or two more passes. I pulled up and to the left and started back toward the leader who had turned right and was heading southeast toward home. As I did, I looked down and could see the helicopter burning.

As I caught up with the lead F-4, I noticed that it looked different. The top of the rudder was fluttering in the five-hundred-mile-an-hour breeze. The metal strips attached just at the top of the stabilators where they meet the vertical stabilizer were bent back and flopping around. The wingtips were fluttering at the trailing edges. Even the streamlined red beacon light located in the leading edge of the vertical stabilizer was broken out. Various inspection panels were popped loose and some were even missing.

The aircraft commander did not look at me. The rearseater was bent over with his head on his left arm, which he had laid along the left canopy rail. He didn't move.

"Lead, that plane has been overstressed. Are you guys okay?" I queried. "Yeah. We're okay," he answered.

His backseater raised his head up and looked at me. He shook his head "No." We were ripping along now at normal cruise speed. Things certainly weren't normal. Panels and wingtips were fluttering. The top of the rudder looked like it was about to depart.

"Lead, don't you think you'd better slow down. Your jet is really bent. Something is going to come off," I suggested.

The aircraft commander slowly turned his head and looked at me. He didn't say a word but in a couple of minutes he pulled the power back and we slowed down somewhat.

We were approaching within radio range of Cam Ranh Bay. He hadn't made any calls.

"Lead do you want me to call and declare your emergency for you?" I asked. I really didn't know what he had in mind. I didn't think he would go in and try to fly a normal pattern without saying anything, but things were already so bizarre that I wasn't sure.

He just turned and looked at me again.

Finally, "Cam Ranh tower. 'Whiskey 99' declaring an emergency. Request straight-in landing."

The crash and rescue squads were standing by when the bent F-4 pulled off the runway in the dearming area.

I went around from final approach after flying a chase position with him and pulled up into a closed traffic pattern for my landing.

It turns out that indeed the lead aircraft commander had been fooled also by the rising terrain behind the target. He had pulled so hard that his Phantom snap rolled coming off the pass. Luckily the F-4 had already started to climb. They came out of the roll still in a climb. I had missed the whole thing because I had looked down in my own cockpit just before it happened.

After they climbed up to the high downwind where I had finally seen them, he had decided to go home, and rightfully so. I can certainly understand his having been deceived by the terrain. I sure was. I do question why he didn't call me off of my first pass or at least warn me. Perhaps there wasn't time. I like to think so.

The backseater went to the hospital. His back was severely strained during the high-G, high-speed, snap roll. He didn't fly for a few weeks and never again with that frontseater.

As I understand it, the backseaters immediately issued an ultimatum to the correct person. "They simply were *not* going to fly with that guy again." (As the old saying went, "What were they going to do if they didn't? Send them to Vietnam?)

They didn't have to fly with him again. He got a *different* job.

Jam lucky

Ho Chi Minh Trail, Laos, 24 February 1966

The word jam can be used in several ways: musical jam session, bread and jam, "tight spot" jam, jam your finger, gun jam. On this particular day, one kind of jam got me out of another kind of jam.

The area we were assigned to strike was particularly hot. Five aircraft had been shot down there the previous day. The North Vietnamese were short of a lot of supplies and modern equipment at times, but they were not short of brains or ingenuity. They could do amazing things with what they had available. Their work ethic was incredible. Where there had been no guns and no threat one day, overnight the place would be crawling with them. This was accomplished at night over extremely rough terrain, sometimes with very few, or no trucks. That's what happened. One day, no threat, the next, five aircraft shot down by heavy antiaircraft fire.

Our job was to try to knock out the guns. Our preflight intelligence briefing had estimated five to seven heavy antiaircraft gun positions controlled by a single director. He would select only one member of an attacking flight and all guns would concentrate their firepower on that aircraft until it was eliminated, then another would be selected. As you can imagine, the odds of hitting an aircraft were greatly increased if all guns were firing at only one plane at a time. Other members of a flight, especially those ahead of the unlucky one, sometimes saw no groundfire at all.

Three was the unlucky number that day. I was flying as the number three aircraft in a flight of four F-4 Phantoms. We were several hundred miles northwest of our home base at Cam Ranh Bay. We were each carrying four seven-hundred-fifty-pound dive bombs and a centerline Gatling gun with over a thousand rounds of twenty-millimeter explosive ammunition. Because of the distance and the aerodynamic drag from the weapons, we expected to have only enough fuel for one dive bomb pass, then haul ass out of there. I sure as hell didn't want to stick around and perhaps join some of the pilots still evading being captured on the ground. They were waiting for the rescue helicopters that were having to wait for us to kill the guns that were keeping them away from their rescue attempts.

The weather over the target area was overcast at around seven thousand feet with visibility three to five miles. Whiskey 11, our flight leader, asked for a fuel and armament check. We each checked in with our fuel remaining and bomb switches hot. After taking spacing from the plane in front of me, I could see the small airfield near where the guns were supposed to be. Just as the lead F-4 rolled in for his one dive bomb pass, the sky around me lit up with flak.

I looked down as I began jinking maneuvers and had absolutely no problem locating the gun positions. When you're looking down the barrels of numerous antiaircraft guns all firing at you, it's easy. Was it scary? Damn betcha! I had no problem seeing how they had shot down those airplanes. I pulled up into the clouds and changed direction trying to throw off their aim. As I rolled out of the clouds inverted and varied my direction and altitude as much as possible, I couldn't believe how many rounds were going off all around us. I also couldn't believe they hadn't hit us.

It was my turn. As I rolled upside down and pulled the nose down to line up on one of the gun positions, I could see where lead had hit one of the targets, and two's bombs were walking right up to another. I placed the "pipper" of the gunsight on the heaviest area of muzzle flashes and adjusted the airspeed as my backseater called out our altitudes.

Thirty-seven-millimeter shells were going off all around us as we roared down the chute. It was incredible that they hadn't hit us. I could even see small vapor trails behind some of the yet unexploded rounds as they flew past the canopy in the moist air. Finally, I rippled all four bombs off and simultaneously pulled and rolled my Phantom, desperately trying to foil the gunners' aim. I looked down during one of the jinks and shouted "*Yes*!" as my bombs found their mark. One of the antiaircraft gun positions became a blazing crater in the blink of an eye. I checked the throttles at full military thrust and started to look for the lead aircraft to the southeast toward home base.

Suddenly the radio crackled with the call, "Whiskey flight, arm the guns, and we'll make one more pass!" Yep, instead of being home free, I still had to pass "Go."

"I'm not believing this," my backseater said. "We've actually got to go back in there." I didn't want to believe it either, but there was Whiskey 11, rolling in for his strafing pass. Not a single shot that I could see was being fired at him. No wonder he was so brave. He hadn't been fired at on his first pass either, I found out later.

In the meantime, we had really pissed them off! They were throwing everything they had left at us. Up in the clouds again we went. I wished I could stay up there. I thought seriously about it for about a second, then back down I came. Several rounds went off so close I could hear them in the cockpit. Pieces of shrapnel pelted the Phantom somewhere. I picked a target and rolled in again. Before I could get in range, several rounds whizzed by the right side of the canopy. I shoved forward on the stick and then pulled up the nose again toward the target. Now in range, I squeezed, BANG! Instead of the buzz of the Gatling gun at one hundred rounds per second, I got, BANG! We were like sitting ducks! No gun, no need to stick around! I immediately rolled left and pulled as hard as I could.

The noise of the explosion and the violent jolt were followed by the most unnecessary radio call I have ever heard, "THREE, YOU'RE HIT!"

I pulled the nose toward the ground and yelled to my backseater that I was all right. I didn't want him to grab the stick, thinking that I was hit. Rounds of antiaircraft shells were still coming our way. I rolled upright and had to pull eight Gs to miss the ground. I rammed in full afterburners, and we passed five hundred and sixty knots and an antiaircraft gun site at the same time. I have a vivid memory of looking to my right, directly at the gunners. They were firing away and frantically trying to crank their guns down low enough to hit us. They couldn't do it. We were almost level with them and now passing at about six hundred knots!

As we approached six-fifty, I pulled the stick back, and we shot up through the clouds in a near-vertical climb. As we punched through the cloud deck, I pulled the throttles out of burner and rolled and pulled into a more normal climb attitude. For the first time, I looked at the engine instruments and started checking for damage to anything besides my shorts. We could see nothing unusual from inside the cockpit. The rest of the flight began joining up on me, and my wingman closed in to check me over.

He happened to be my squadron commander and the author of that radio call I mentioned. He said the only thing out of the ordinary that he could see was that my right wing's three-hundred-and-seventy-gallon drop tank had exploded. The rear section had been blown off, and the nose of the tank had blown outward and split. It was bent backwards over the tank itself. He also reported several small holes in the bottom of the right wing, but nothing was leaking. Evidently they had not hit any vital parts, and everything else appeared normal.

Later, during the debriefing of the mission, we discovered that the first two Phantoms didn't think that they had been targeted. In fact, they had trouble locating our targets without the muzzle flashes. The leader didn't realize that I was being hammered although groundfire was reported. My wingman hadn't been fired at either, as far as he knew.

Funny how luck works. Remember at the beginning, I said that a jam got me out of a jam? If you take a model of an F-4, look at it head-on, then rotate it around its longitudinal axis (like it is turning hard), the drop tank makes an arc through where the canopy previously was. I believe that if my gun hadn't jammed, causing me to break left immediately, I wouldn't be sitting here writing this account. I think that the shell that hit my drop tank would have come right through the windscreen: hence, the title of this story, "Jam Lucky."

Four months later, in June 1966, my right fuel drop tank was blown off the airplane. But that's another story.

Parts is parts!

Cam Ranh Bay Air Base, RVN, 1966

All through my Air Force career I have heard that "If you've never landed on an aircraft carrier, you ain't s___!"

Of course it was Naval "Aviators" talking. I confess to you guys that I always wanted to land on a carrier. (I *know* if *you* can do it, then with the training, so could *I*.) However, I will not concede that landing on a deck necessarily makes you better in the air. I *will* acknowledge that it does cause your task to be a lot harder when your "air" time is over, and I salute you for developing and maintaining that skill.

I have always wanted to go out and watch "Carrier Ops." That has to be one of the most fascinating and intense operations ever devised by man. Navy personnel are to be congratulated and admired for making it work so well. It almost looks impossible.

Some old Navy F-4 "jocks" may correct me, but I think that an F-4 landing on a carrier rolled out only something like ninety feet after hooking the cable. It might have been even less. An Air Force F-4 engaging the barrier (taking the cable) stopped in about nine hundred feet or so. I have hooked a cable a couple of times, and I can tell you that it's quite a sudden stop. You had better have your shoulder harness tight and locked. Even then it feels like your head is going to keep going. I can't imagine what stopping on the deck in ten percent of that distance must be like. Maybe that's why a lot of Navy aviators have necks like bulls. It could be an occupational necessity.

Just south of Cam Ranh Bay a few miles, the Air Force established another Air Base, Phan Rang. I believe Phan Rang had an F-100 unit for a short time before some F-4C squadrons were brought in.

One day I noticed an additional "occupant" on our ramp. It was an F-4 with no landing gear. It was sitting on its belly near the east edge of the ramp. It had been brought up from Phan Rang, and it was going to be shipped back to McDonnell to be rebuilt.

The word I got, "it" had happened during a maintenance test flight. The pilot was a field grader (major or above) with a key staff position in the wing. During the checkout flight, he had experienced a utility system hydraulic failure. That meant the loss of several useful items, such as normal brakes for instance.

The recommended procedure in an F-4 with utility hydraulic failure was to engage the approach end arresting barrier, if one was available, with your tailhook. That way you avoided steering and braking problems after touchdown. The cable would stop you essentially straight down the runway if you engaged it near the center as you were supposed to. The arresting cable was usually located about fifteen hundred feet or so from the approach end of the runway. This gave you plenty of room to get the jet down before reaching it. (You Naval aviators are probably saying, "That's about thirteen or fourteen hundred feet more than you need.")

Normally the pilot's biggest concern was that of the tailhook bouncing over the cable. If that happened, the pilot had to try to stop with the emergency brakes or go around for another try. (The second choice is the Navy's *only* choice.)

However! There is another possibility. On a carrier it will probably kill you. On a runway, it might.

Phan Rang's runway was like Cam Ranh's. It was about ten thousand feet of aluminum laid out on the ground. There were no "overruns," which would have been hard-surfaced areas about a thousand feet long at both ends of the runway. There was just the dirt and then the edge of the aluminum.

The Phantom maintenance test pilot had called in and declared his emergency situation as required. The crash and rescue equipment had responded and was standing by. The mobile control jeep was out by the runway with an F-4 pilot manning it to assist if necessary.

The slightly disabled F-4 was turning onto a long straight-in final approach. Everything looked good. He arrived at the normal descent point to maintain a proper glidepath to the touchdown. Everything still looked good.

Then to the F-4 expert in the jeep it began to look like the Phantom might be dropping just a little bit below a desired glidepath: not so good. He waited a short time thinking that the fighter pilot would lessen his descent rate slightly to get back on the proper profile. The pilot didn't correct.

The mobile control radio crackled to life. "Your glideslope looks too low. Please check it." There was no response either on the radio or in the Phantom's descent profile.

"Please check your descent rate. You're going low on the glidepath," called the F-4 guy in the mobile jeep.

"Mobile, I'm flying this airplane, not you!" came the field-grader's terse reply. (I heard that the mobile officer was a mere captain.)

"Yes, Sir. It's all yours, Sir," came the crisp response from mobile.

The F-4's exhaust began kicking up clouds of dirt as it approached the end of the landing strip. As the aircraft passed over the end of the runway everything made it, except the tailhook. It hooked the edge of the ten-thousand-foot-long one-hundred-foot-wide aluminum runway. The runway and the tailhook didn't budge. The tailhook stopped momentarily along with a substantial portion of the structure that had once attached it to the jet. The Phantom kept on going but leaving some of its "vitals" behind.

The hapless hookless F-4 sat on our ramp for a long time. I noticed that it seemed to be getting smaller and smaller every time I went by it. First a canopy, then a rudder, then a wingtip, and so it went. The once proud jet started looking like a dead bird being picked over by vultures. When it finally left for St. Louis, most of it was still flying combat missions within dozens of other F-4s.

"Parts *is* parts," you know!

Come on baby, we can do it!

Laotian Mountains, 1966

I was one of those pilots who thought that if *anyone* could do anything in an F-4, then so could I, *and* do it better! I realize that sounds braggadocios, but is it really, if it could be backed up by performance? Maybe. Maybe not.

Fighter pilots know what I'm talking about. In my opinion, you're probably not a very good fighter pilot if you don't feel that way. It was said to me once that "If you didn't have a pretty strong ego, you probably shouldn't even get in one of those damn things!"

There were only two of us. We had been scrambled off an alert status to bomb some trucks that a forward air controller had spotted. He wanted someone to respond quickly before the enemy had a chance to move them to another hidden truck-park or get completely away.

I always liked flying off an alert scramble because the targets were usually better than the ones coming in on the target frag orders daily. Either the Army needed quick assistance or a FAC had spotted something "hot" as in this instance. Also, it was challenging to go into a target area that you didn't know much about. It was especially tricky if it was in the mountains or if there was some weather thrown in for good measure. It made for some interesting situations.

We had hammered the truck-park and were rewarded with several explosions as some of the trucks and their cargo blew. There hadn't been any substantial groundfire in return, so it had been pretty easy. It was fun because it was in a valley between some very high ridges, and you really felt some great sensations of your speed in such terrain.

As we were dropping our last bombs, the forward air controller asked if we had enough fuel for one more target that was nearby. We did, and he told us to arm up our guns and to follow him.

We couldn't *really* follow him because of our tremendous speed difference, but we "S" turned and circled over him as he made his way toward the northwest. He had only gone two or three miles when he turned right and headed up a deep dead-end valley. We couldn't really tell how deep it was from our altitude as we circled and watched him. We could catch glimpses of what appeared to be a trail or narrow road winding under the trees. At the far end of the valley was a near-vertical cliff. It rose straight up out of the thick forest for what appeared to be around fifteen hundred feet. The FAC flew right toward the sheer face of the rock wall and fired a smoke rocket. He then turned tightly and came back out toward the mouth of the dead end valley.

"Your target is ten meters to the left of my smoke at the base of the cliff. It is a large camouflaged cave with a suspected ammunition storage area inside. If you can throw a few rounds of 20 mike mike (20 millimeter) in there, we might just light em up," the forward air controller said.

"How tall is that cliff above the cave," I asked.

"Looks around a couple of thousand," the FAC answered.

That jived with what I had thought, so I was fairly satisfied with the "picture." I rolled in at the mouth of the valley and roared up it in near level flight. I did not want to be in any descent toward the bottom of the cliff. I figured that this was going to be tight enough without making it worse, but remember, if anyone could do it, I could.

I set the speed at four hundred and sixty knots and headed just to the left of the smoke at the base of the cliff. I couldn't see the cave because of its camouflage, but I knew approximately where it should be from the FAC's smoke rocket. I kept moving my eyes from the base of the cliff to the top and back. I knew that I was going to have to fire somewhat out of range to make the pull and adjusted my aim point higher to allow for some drop of the high-explosive cannon rounds. The cliff was looking taller and taller by the second.

I squeezed off a three-second burst and immediately began pulling hard on the control stick. I had also gone to full afterburners. It was looking really "tight." I pulled harder still. The Phantom was shuddering with its effort to make the vertical turn. The airspeed was deteriorating rapidly. The wall was getting closer and closer. I eased the stick back farther and farther. The cliff must have been twice as high as it had appeared. All I could do was try not to pull too hard to give the big F-4 a fighting chance to save us.

I saw the top edge of the rocks drop away just a few feet beyond the belly of the fighter. I released some back pressure and looked at the airspeed indicator. We were just under three hundred knots. I had used one hundred and sixty knots plus of energy making the pull.

"Abort Two! Abort your pass immediately and pull off high and dry!" I yelled at my wingman who I knew was on his way up the valley. "We *ain't* doin' this!"

As two joined up on my wing, I briefed the forward air controller on my problem on the pull-off and told him to get someone with a Bullpup missile or to forget it. (A Bullpup is a missile that can be fired from a fighter from a long distance and accurately guided to the target with a little control stick in the cockpit.) I told him that the cliff must be a lot higher than we had thought because we had almost busted our ass about three-fourths of the way up it.

As we headed back to Cam Ranh Bay still in one piece, I affectionately patted the F-4 on her glareshield: "I *knew* we could do it!"

The Silver Star, posthumously

Cam Ranh Bay Air Base, RVN, 1966

My flight leader and I had been scrambled from runway alert to try to knock out some automatic heavy weapons that had a U.S. Army special forces camp pinned down in the central highlands. The camp had taken heavy casualties and needed help to defend themselves and so they could get choppers in to evacuate the wounded.

When we arrived over the target area a few minutes later, I told my backseat pilot that because of the low-altitude weapons we were carrying, this was going to be an extremely dangerous situation. We had been trained to always try to avoid a duel with heavy automatic weapons at low altitudes. The bombs we were carrying were designed to be dropped at an altitude of forty feet. The only protection we possessed was our speed, but on this mission, we couldn't exceed four hundred and sixty knots (about five hundred and thirty miles per hour) because of the speed restrictions on our bomb deliveries.

Because of the dire situation in the special forces camp, we decided to attack anyway. The ground fire was extremely intense on our first pass, and my aircraft was hit, however not seriously. I told my backseater that this was so hairy that one of us was going to get shot down. Sadly, that proved to be a prophetic statement. On the very next pass, as I was on the downwind leg of my bomb pattern, my leader's aircraft exploded into a fireball as he pulled off of his bombing run.

I knew that the enemy gunners would shoot the pilots in their parachutes if they had to bail out too close to the guns. I immediately rolled in toward the guns from where I was to try to draw the enemy's attention from my leader's plane. As I had hoped, when I began firing my cannon, the gunners turned all their attention and guns toward me. I rolled upside down, pulling out just above the trees to avoid their bullets, and started trying to catch my leader's F-4. All I could see from behind it was the wingtips and the top of the tail jutting out of the raging fire.

I couldn't believe the Phantom was still flying and started yelling for the pilots to bail out! They were far enough away from the guns now, and their F-4 couldn't possibly fly much longer. I saw a canopy eject from the airplane and an ejection seat tumbled by my right side with one of the pilots. His parachute opened immediately, and I continued to catch up with the burning plane.

As I joined up on my leader's left wing, I could see him struggling to get the front canopy off. Evidently the explosion and fire had jammed it, and it would not open or jettison. Unlike some other fighters I've flown, an F-4 would not allow the pilot to eject through the canopy because of its ejection system design. Consequently, my friend was trapped in the burning plane.

He looked at me with sad brown eyes. I was close enough to see their expression. He then shrugged his shoulders like I'd seen him do dozens of times on the ground. Then he saluted! His Phantom rolled suddenly toward me out of control. I pulled up sharply to avoid it. It impacted inverted, about one thousand feet below me with several bombs still aboard. A millisecond later, there was nothing but memories and dreams.

My wife tells me that if she were Joe's wife and three children, she and the family would want to know what his last mission was like and what happened to him. The question of whether to contact them has bothered me for over twenty-nine years now. I don't want to open up old wounds and hurts; however, if I thought that they wanted to know what happened that day, and that it would help, I would be willing to tell them.

Joe was a true hero that day. He was a calm, brave man when facing life's greatest challenge. I only hope that I can meet that final challenge as well, when it comes.

He was awarded the Silver Star, posthumously.

Ladies and gentlemen, on your left . . .

Coastal Plains, RVN, 1966

The forward air controller had received a lot of small arms fire from the area the day before as he passed over. He had marked his map and gone back to his airstrip just a few clicks (nautical miles) away. He had been low on gas and decided to call it a day as it was getting toward dark anyway.

The next morning our "frag order" (target location and other pertinent information) just said "enemy troop concentration." Our Phantoms had been armed with napalm bombs along with a full load of incendiary twenty-millimeter shells for our Vulcan cannons (a Gatling gun mounted on the lower middle of the fuselage). The target wasn't very far away to the south, so our takeoff time was only about twenty minutes before our target time. I was only scheduled to fly once that day, and I was looking forward to a relaxing afternoon of volleyball and a nap. It looked like an easy mission.

I was leading the flight of four, and I checked them in on the forward air controller's frequency, "Whiskey 21 check," I called. "Two, Three, Four," they quickly replied in turn. It was a beautiful day near the coast of the South China Sea. I wondered if our enemy troops were still a concentration and whether they were enjoying the day, so far.

"Whiskey 21, good morning," said our FAC immediately after I called him. "I've got you in sight if you're the F-4 four-ship just north of the target area," he continued. F-4s were easier to spot than most other jet aircraft because of the trails of black smoke pouring out of the tailpipes. The only time the engines didn't smoke was when they were in afterburner.

"That's us," I answered. "We're just descending through six thousand feet for five."

"Okay, here's the poop. The bad guys didn't disappoint me. I just took a little excursion past the place where they took the pot shots yesterday, and they couldn't resist taking a few more. I want you to make them real sorry for trying to hurt me. Hang on a second, and I'll mark the target. I'm just to the east of it, over the beach," he explained.

"I've got you in sight. Waiting for the smoke," I called.

"Rocket away," the FAC called as I watched the little 2.75-inch rocket dart from under his wing. He immediately banked hard away from the target area and back toward the beach. He then banked steeply back to the north and looked at where his rocket had landed. "Bingo. I may not even need you guys. That was a bull's eye," he laughed. (He was joking because all his rockets were designed to do was make smoke when they hit.) "Go spoil their breakfast, Whiskeys. You can roll in from where you are now and make your runs basically southeast to northwest. I'll be orbiting just offshore at three thousand feet."

"Whiskey lead's in hot on the smoke," I said as I rolled in from the southeast. As I approached the target, I could see muzzle flashes from what looked to be thirty-caliber shoulder-fired stuff. If they had anything bigger, they hadn't had time to set it up yet. I thought to myself, "They'd better hurry." I dropped and pulled. For some weird reason, I always really enjoyed the sensation of pulling Gs off a target and then banking hard to get up on a "downwind" leg for the next pass. I leveled off and watched as the other three Phantoms in my flight plastered the target area all around the original rocket smoke.

The forward air controller wasn't saying anything, so I assumed he was satisfied with our bomb coverage. Four said he hadn't seen any ground fire on his pass when I asked him. I told the FAC that it looked to me like just one more pass would probably suffice and asked if he concurred. He said one more should do it, so I set up my bomb switches to drop all of my remaining napalm. No gun passes would be necessary, at least not on this target. The target touched the bomb sight. I "pickled and pulled" (dropped the bombs and pulled back on the control stick).

"Whiskey look out! There's a. . . !" was all I heard before I saw it. The French-built Caravelle jet airliner was one of the prettiest aircraft ever built, but not from the angle that I was seeing it. Its belly was completely filling my windscreen. It was only about two thousand feet above our burning target and slightly northwest. It appeared to be circling in about a thirty-five-degree banked turn to the left. I snapped my jet completely upside down and pulled for all I was worth. The inverted Phantom was shuddering and shaking in as hard a change of direction as the airspeed would permit.

I have no idea how close we came, but someone in the flight told me later that the airliner and my F-4 appeared to merge. I pulled until my jet's nose was well down before I rolled upright again. I looked up and back, and the Caravelle was still moseying along in his left turn. My backseater had not even seen the airliner before the near collision. He had been looking back at the target.

He had been doing his job and keeping the other flight members in sight for us. The airline pilot had not been doing his. I don't know what in the world the Caravelle pilot could have been thinking. Not much would be my guess. He sure wasn't keeping his passengers safe, which should be an airline pilot's primary duty. I imagined the Caravelle captain on his public-address system: "Ladies and gentlemen, on your left you can see some United States Air Force jet fighters on an airstrike. . . ."

As we climbed toward the north out of the target area, the forward air controller gave us our BDA (battle damage assessment). When my backseater had finished copying it down, I asked the FAC, "Do you still have that Caravelle in sight?"

"Yep. Believe it or not, he's still circling the area. Maybe he's lost," the forward air controller half laughed.

I thought to myself, "If he's not, he almost was, and everyone else with him." I don't think he ever had a clue. What's that old saying, "Ignorance is bliss?"

Next time, use the bomb!

Cam Ranh Bay Air Base, RVN, 1966

In the early days of the F-4 program in the United States Air Force, things were done quite differently than later on. As I related in a previous story, a certain Air Force general had decided that instead of navigators or radar intercept officers in the rear seat, there should be a pilot. His theory was that with two pilots and two big engines, there would be no more F-4 accidents. Right! Consequently our GIBs, or "guys in back," were brand-new pilots, fresh out of pilot training, who were then trained to be radar-intercept officers. That's not exactly what a new pilot really wants to do. Keep this point in mind while reading this story.

Due to various reasons, one of which was combat losses, it was decided that several backseaters would be upgraded to aircraft commanders, or frontseaters. Of course, these individuals were very excited about their good fortune. They were also determined to prove that they were true tigers, worthy of their selection.

I was elected to be one of the "lucky" instructor pilots who would check out these new aircraft commanders. Our upgrade program was short. One transition ride to familiarize them with the front seat, then, believe it or not, combat. Yep! Transition two was as number four in a flight of four on an actual combat mission. After several of these baptisms of fire, we were to cut them loose with a regular backseater.

Now the Phantom had very poor visibility from the rear seat, particularly looking forward over the frontseat pilot's shoulder. Especially if the frontseater was about six-foot-three weighing around two-thirty!

Our assigned target that day was a group of buildings suspected to house enemy communications, billeting, and, possibly, arms and ammunition. Our particular F-4 was armed with a twenty-millimeter Gatling gun and napalm. After locating the target and taking spacing on the rest of our flight, we began our first run-in for a napalm drop.

The terrain was hilly and covered with numerous palm trees. That required us to make the bombing passes in a slight descent. We normally dropped napalm between forty and forty-two feet of altitude, at an airspeed of four hundred and sixty knots, about five hundred and thirty miles per hour. This equates to around eight hundred feet per second across the ground. In other words, we were flying at 1/20 of a second *above* the ground. As you can readily see, there was not a lot of decision making time!

I strained to see over my student's left shoulder as palm trees ripped past. I ascertained that he was lined up with one of the buildings. It looked like a good run. I was guarding the control stick in the rear seat, but not touching it. I didn't want to affect the student's feel for the airplane, but I stayed close, just in case. As the building rushed directly at us, I waited for the "pickle and pull," as we called it. Time was up! I yelled "Pull!" as I grabbed the stick and yanked back. We cleared the building by mere inches.

When I felt the frontseater also applying back pressure, I relaxed my grip and used a few choice words describing his ancestry and inquired about any recent death wishes that I might need to know about. He assured me that he hadn't lost his mind and would do better on the next pass. As we achieved lineup on another structure, I again tried to stretch my neck to see better. No way, though. Those fine folks in St. Louis built a great airplane, but obviously didn't think seeing out of it was a priority.

Again, I guarded the stick. He surely wouldn't do it again, I thought. *Wrong!* This time I was a fraction too late. As we both pulled, I felt the new sensation of knocking the roof off a building with an F-4. At that speed, it didn't last long.

We were extremely fortunate that the local contractors mainly used thin wood and straw in their roof construction. We were also extremely fortunate that the Phantom is such a fantastically strong airplane. The frontseater was extremely fortunate that he outweighed me by about forty pounds! Looking back over my shoulder, I could see that the napalm had been wasted. We had knocked off the roof and part of the second floor with the airplane.

After picking straw and splinters out of some of the belly seams, and excepting a few minor dents, our ground crew declared our five-hundred-and-thirty-mile-an-hour battering ram again airworthy.

After a very caustic and extensive debriefing, I declared the student conditionally airworthy, as long as it was with another instructor!

Evidently he had learned his lesson and completed his upgrade training successfully. He also finished his combat tour in Vietnam successfully. Me? I picked a smaller student.

By the way, I heard he went on to fly for a major airline after his tour in the Air Force. I'm sure that he is a senior captain by now. Want to know which airline? *Nah!*

What tree?

Cam Ranh Bay Air Base, RVN, 1966

Another incident involving a pilot in our fighter wing at Cam Ranh Bay served to again demonstrate the inherent structural integrity of the McDonnell F-4 Phantom II.

The weapons were napalm bombs and twenty-millimeter cannons. The target was North Vietnamese Army trucks located by a forward air controller in one of their hidden "truck parks." The F-4 flight had been diverted from their primary target for which their napalm was more suited, but the FAC did not want the NVA to get away to another hidden location with the trucks that he had spotted; therefore, the napalm would have to do. If it blanketed the trucks and their cargo, it could do the job.

The area was heavily forested. Tire ruts could be seen leading into the trees, but at F-4 speeds, the trucks were extremely difficult to visually acquire; therefore, the four F-4s were more or less trying to cover the whole area. The terrain was rugged, requiring the Phantoms to modify their bombing runs. Normally napalm was dropped in level or near-level flight at an altitude of forty feet above the ground. The hills and mountains surrounding this particular target would not allow the preferred method.

The F-4s were really getting down "in the weeds" as they bottomed out of their descending napalm passes. The bombsight was either ignored or barely utilized in such terrain. To concentrate too much attention on it and not enough outside the cockpit could be disastrous. Most of us had gotten so accurate with napalm that we didn't use the bombsight anyway. Of course we didn't use forty feet as our minimum altitude either. We just aimed the Phantom at the objective. When we reached the point where we would either have to fly the jet right through the target or pull up, we would hit the bomb release button and pull. We seldom missed.

The Phantom pilot released his last napalm bomb and pulled. As he did so, he felt a slight jolt. The heat from his and the other F-4's napalm had caused so much turbulence that he didn't think much about it, that it was just rough swirling air.

The flight joined up for their return trip "home" to Cam Ranh Bay. The wingman on the right noticed it first. The tail of the subject Phantom had a large hole punched through it. As he moved in for a closer look, he was astonished. The right wing root was yanked away from its attach point on the lower engine intake. A huge gash was rammed into the leading edge of the wing several inches wide and many inches deep. Most of the right wing area of the jet had peripheral damage. Even with all this, the pilot of the F-4 hadn't noticed that anything was amiss in the way the Phantom handled.

The fighter had hit a tree approximately ten inches in diameter. The tree had rammed back into the wing until it met with the main spar. It broke off and proceeded to punch the big hole through the tail section of the Phantom. The normal wing sweep angle on an F-4, if I remember correctly, was forty-six degrees. The "tree-modified" wing now had an effective sweep of fifty degrees or more, and it had flown just fine.

With this incident, another chapter was added to the rapidly expanding saga of the ruggedness of the McDonnell F-4. Is it any wonder that pilots fortunate enough to fly the Phantom II loved her?

The race is on

Cam Ranh Bay Air Base, RVN, 1966

The term "monsoon rain" sounds very exotic and mysterious. In the United States, they would just be called thunderstorms. They could be extremely heavy at times, especially during Vietnam's monsoon season.

Recall that the runway at Cam Ranh was aluminum. When we first started operating from it, there was an antiskid surface applied to it that worked well, except when it was covered by heavy rain. Also, after months of flight operations, the antiskid surface began to wear off and added to any potential stopping problems.

I was the number-three aircraft in a flight of four that particular day. It was during the season for monsoon storms, and we had already experienced several huge examples recently. As we approached from the northwest, we could see several storm cells nearing our base and were hotfooting it back to try to beat them. Hurrying down initial approach at three hundred knots, we could see a solid-looking wall of water just north and west of the far end of the single runway. We were flying in tight echelon formation and approaching the end of the runway for our overhead pattern. We were taking spacing of three seconds between aircraft before rolling into our nearly ninety-degree-banked three-G turns to downwind where we would lower the gear and flaps for landing. This overhead, or fighter break, was the quickest method to get all flight members on the ground. We usually landed from this pattern.

As I turned from the downwind onto the base leg of the pattern, I could see the lead aircraft on short final about to touch down. The rainstorm was just crossing the north, or far end, of the runway. As I reached a point about ninety degrees from the runway heading, I saw the lead aircraft disappear into the water wall. The Number Two aircraft in front of me had just touched down, and now the rain was about one-third of the way down the runway. He entered the rain just as I was about to touch down.

The storm was kicking up the wind, and I pulled my drag chute handle as I touched. Halfway down the runway, I entered the heavy rain. Visibility dropped rapidly, but was still probably a half mile. I touched the brake pedals when I had slowed enough, and the antiskid system began cycling furiously as the tires tried to grip the slick aluminum surface. I was to the left of the center of the one-hundred-foot-wide runway. (Each plane alternated runway sides with the aircraft in front.) The antiskid was still trying to find some traction when an agitated radio call penetrated my headset. "Four's just lost his chute!" I looked in my right-hand rearview mirror, and there he came, closing the distance between us.

I was still a couple of thousand feet from the arresting cable, which was located about fifteen hundred feet from the far end of the runway. If an F-4 had any difficulties stopping caused by a loss of brakes, or as in this case, a loss of its dragchute combined with a wet runway, a tailhook on the rear of the aircraft between the engine afterburners could be lowered. It was designed to hook the cable, which was suspended a few inches above the runway. The cable then rapidly slowed and stopped the plane in about nine hundred feet. Unfortunately, it did one other thing. It lifted the cable several feet above the runway's surface.

I quickly visualized what would happen if Number Four passed me on the narrow runway, hooked the cable, and raised it in front of me. The resulting impact and fireball as our Phantoms collided, following my nosegear impacting the raised cable, would be horrendous. I had no choice. I *had* to outrun him to the cable. I jammed the throttles to military power (full power without afterburners). I glanced in the mirror, and Four's F-4 was rapidly overtaking me. My engines responded, and I began to accelerate toward the arresting cable. His nose radome was visible beside me now. My Phantom continued to accelerate as the thrust took effect. His nose began dropping back out of my peripheral vision.

As my racing fighter passed over the arresting cable, I jerked the throttles to idle and pushed on the brake pedals for all I was worth. It felt like I was trying to stop the world from turning.

Now, if Number Four's tailhook didn't skip over the cable (always a possibility), we might still come out of this alive. "Four's got the cable!" my backseater said excitedly as he watched over his shoulder. Now at least I knew that Four wouldn't be ramming us from behind.

I was pushing on the brake pedals with all my strength, trying to stop in the short remaining distance. The antiskid system continued its incessant cycling. I could see the end of the runway rapidly advancing with the deep sand beyond it. Just then, I felt the Phantom's tires grab at the runway surface. Some hope returned. Maybe I could still extricate us from this impending disaster. Until then, if unable to stop, I had intended to head straight off the end of the runway into the sand and suffer the intensely undesirable consequences. (At the very *least*, it would have ripped the landing gear from under my F-4. Previous unwanted excursions off the runway by other Phantom pilots had already proven that. At the worst, we could cartwheel upside down and be trapped in our burning airplane.)

Number Two had almost cleared the narrow taxiway leading from the runway's end into the dearming area. I pushed the microphone button and yelled "Move it Two!" He was listening. I saw his engine eyelids close down as he cobbed his throttles. I stood on the right brake and engaged the nosewheel steering. Two had exited the narrow taxiway and was well clear as I entered it in a skid. I thought for a moment, "I am going to lose it," but then regained control as the nose came back to the left.

I finally got the Phantom stopped. The nose radome was almost hanging over the sand on the northeast edge of the dearming area. I was shaking like a leaf as the adrenalin pumped. I looked back to my left and could see Four's Phantom stopped on the runway with the cable jammed in the tailhook. My backseater then began reading to me the "After-Landing" checklist. I couldn't hear a word he said. The blood was pounding too loudly in my ears.

I learned later that there had been nothing wrong with Four's dragchute. He had failed to rotate the handle far enough to the rear to lock it in the deployed position. When he had let it go, it had simply rotated forward and released the dragchute as it was designed to do. I felt like doing a little loud pounding on *his* ears.

It was hard for me to believe that we had escaped from that intensely ominous scenario without a scratch. There were four or five outstanding opportunities for a major accident as we had "skated" down that wet runway, and we had avoided them all!

Small world, isn't it?

Cam Ranh Bay Air Base, RVN, 1966

The Pacific's monsoon season sometimes affected our flight operations drastically. It seemed that when we needed it the most, our ground-controlled radar equipment tended to head south. Our GCA controllers were among the best anywhere, but they were hampered by their equipment's limitations during heavy rain.

I was in my room in the hootch, as we called our barracks. The rain was pouring down in sheets when I heard the roar of F-4 engines. I was surprised that anyone was up flying in that terrible weather. From the sound of it, he had just executed a missed approach from a landing attempt. It also sounded like he had been lined up a long way to the east of the runway. In fact, he seemed to be over the area of our housing, which was quite a few thousand feet from the runway. I didn't think much more about it at the moment, as I assumed he would divert and land at Phan Rang Air Base, which was about twenty-five miles south of us. Normally the thunderstorms were fairly isolated, which left one or the other of the two F-4 bases clear for landings. Slowly, the sound of those engines faded in the distance, and I continued whatever I had been doing.

There it was again. The roar of the engines was nearer the runway this time, but it was still obvious that he had missed again. The engines' noise filled the building, then just as suddenly, it was quiet. It didn't register for a few seconds what I was hearing (actually, not hearing). Then I heard something else: two explosions very close together. It was the sound of ejection seats!

I grabbed my raincoat and started running toward the flightline. I was joined by several others who had heard the same thing. We also heard the sound of sirens in the distance. They seemed to be just across the runway area to the south and west. We couldn't see anything because of the heavy rain, so we headed to the flight operations building to get the news.

It seems one of our new pilot upgrades and his backseater had returned from a mission with the other members of his flight. His was the last aircraft of the flight to accomplish a GCA. The other flight members had made it down just before the rain's intensity increased.

As the new aircraft commander had broken out of the clouds and spotted the runway, he was too far to the right of it to line up and land from that approach. He executed a missed approach, as he should have, and announced that he was heading for Phan Rang Air Base to land.

A certain colonel, who will go unnamed, activated his radio and instructed the young pilot to make another GCA and land at Cam Ranh Bay. The young lieutenant was evidently swayed by the rank and position of the individual and attempted to comply, although he was getting dangerously low on fuel. On his next approach, the young pilot had again been lined up too far to the right of the runway to land. Again he went around, only this time both engines flamed out from fuel starvation and both crewmembers immediately ejected.

It was just in time. The backseater, who ejected first, swung once in his chute and entered the shallow water. When the aircraft commander's chute blossomed and straightened his body, his feet were in the water.

The aircraft commander and his backseater received very minor injuries and soon were back in action. Of course an accident board was held and the pilot charged with pilot error for not diverting to Phan Rang and landing. Isn't that enlightening?

It certainly was for me!

There are *three* postscripts to this story.

PS 1: It was the third ejection from an F-4 in Vietnam for the backseater. He never suffered anything other than scratches and bruises; however, while on R and R following this particular bailout, he wrecked a rented motorcycle and ended up in several bandages and a cast.

PS 2: About a year later, I was being medically tested for a pilot's job with Pan American Airways in Miami, Florida. There were eight or ten applicants, including myself, sitting in a room while awaiting our glucose tolerance testing. The applicant sitting on my left introduced himself and asked me about my flying background. I introduced myself and told him Air Force. He asked me what I flew.

"F-4s."

He asked, "Did you do any flying in Vietnam?" I replied that I did.

"Where?" I told him "Cam Ranh Bay."

He then said, "What a small world." He stated that he was an Army pilot based at Nha Trang at the same time. Nha Trang was just a few miles up the coast, north of Cam Ranh Bay. He next said, "Tell me, who *was* that stupid S.O.B. who ran that F-4 out of fuel and bailed out of it just before it landed in the bay?"

I looked the Army pilot in the eye and said "Him" as I pointed with my thumb to the applicant sitting on my immediate right.

I said "It *is* a small world, isn't it?"

PS 3: The voice on the radio went on to retire from the Air Force with several stars on his shoulders. I've learned that we just have to accept that sometimes there isn't any justice, at least not in this world.

Rank has its problems also

Cam Ranh Bay Air Base, RVN, 1966

My youngest daughter Tiffany (born in 1969) used to ask me when she was a little girl, "Daddy, do you know everything?" That question would arise after she had asked me another question that was puzzling her child's mind and I would answer it (some of the time). I'm sure it appeared to her that I was really smart because I knew the answer to her question.

When I was an Aviation Cadet, I thought that even a second lieutenant must have most of the answers because he had been able to progress to such a "lofty" position. My career up to that point had gradually dispelled that false impression. My year in Vietnam helped to accelerate that educational process. The previous story is a prime example of the other side of the coin. It taught me that no one, regardless of rank or position, knows everything. It showed me that one of the most dangerous things in this world is someone who thinks that they do. It also taught me that blind obedience to someone just because he had an oak leaf or an eagle on his shoulder is not always the right thing to do. It taught me, if I *knew* I was right, to have the courage to say so and to take any action necessary if warranted. I'm not following *anyone* off a cliff.

Whiskey 34 was my callsign that day. Our flight leader, Whiskey 31, was a visiting lieutenant colonel from one of our higher headquarters. The weather at our primary target was in the weeds. We proceeded to our secondary target, and it was no better. Because we were carrying seven-hundred-and-fifty-pound bombs and Gatling guns, our fuel consumption was very high. Finally, our flight leader gave up trying to locate a hole to get down through the clouds, and we started our trip back from Laos to Cam Ranh Bay. I was relieved that the weather had not cooperated with a hole near the target. There was almost nothing more dangerous than diving through a hole in the clouds over mountainous terrain that was unfamiliar. We weren't even sure of the accuracy of our maps as far as elevations were concerned. Even if they had been accurate, you have to know exactly where you were for them to help. That wasn't always possible when you were circling and descending away from the immediate vicinity of your target.

We were near Pleiku when the call came from my flight leader. He had left our radio frequency for a moment, I had assumed to call for a weather update at Cam Ranh Bay. When he returned to our channel, he had us all go to a new frequency and check in. After we had done so, he began talking to an airborne command post about a possible target. They briefed him on a target near Da Nang. I glanced down at my fuel remaining and began mentally calculating distances and fuel consumption. Something wasn't computing. At our present burn rate, we didn't have enough fuel, in my estimation, to fly to the target area and then to Cam Ranh Bay, even without descending and striking a target. I thought, "There must be something that I'm doing wrong in my calculations, something I'm missing."

The Number Four aircraft in a formation usually burns slightly more than the rest of his flight members. It's because being Number Four requires more throttle movements to maintain position, and he is usually on the outside of any turns. This is especially true in route or spread formation.

I requested a fuel check. It was readily accomplished. My fuel state wasn't even the lowest in the flight, so that wasn't the answer. I had also thought that if I asked for the fuel check, it might cause 31 to recalculate our fuel remaining requirements.

We had turned sixty or seventy degrees to our left and were heading toward the target area. I told my backseater to get out his calculator and refigure our fuel. He had already been doing it because he was concerned also. When he had finished, he gave me the news. We would flame out well before getting back to Cam Ranh Bay if we continued to the newly proposed target. There had to be something I didn't know!

"Whiskey 31, this is 34," I called.

"Go ahead, 34," lead answered.

"Sir, are we going to land at Da Nang Air Base for fuel after the strike?" I asked.

"Negative Four, we're not," he answered quite brusquely.

I thought it over for a few moments before I said, "Sir, 34 does not have enough fuel to proceed to the target and then return to Cam Ranh Bay."

He replied immediately, "Four, I'm leading this flight, and you have as much fuel as anyone else. Stay off the radio."

I couldn't believe that this was happening. I had to be missing something! I asked my backseater if he was sure about our fuel computations. He assured me that he had just checked it again, and they were correct. We were heading about seventy degrees from a direct course to Cam Ranh Bay. My mind was in turmoil. I had been taught to follow orders and to never leave a formation without permission.

My stomach knotted up as I made my next radio call. "31, 34 is leaving the formation and returning to home base," my shaky voice said. Simultaneously, I began a right turn toward Cam Ranh Bay.

The agitated response was immediate. "Four, get back in formation and stay there!"

"No sir," I said and continued turning. He said no more.

All the way to Cam Ranh Bay, my self-doubt gnawed away at me. I must have refigured my fuel a dozen times. It always came out the same. It was going to be so close that I elected to land from a long descending straight-in. I didn't have the fuel for a normal overhead break and landing pattern. I touched down with twelve hundred pounds of fuel remaining. We were supposed to always plan to return to the pattern with three thousand pounds.

As I stepped slowly down on the aluminum matting beneath my plane, I looked up and saw them. There were three F-4s in trail for straight-in approaches. I watched them land and roll out down the runway. Two of them made it into the fuel pits where I waited. The third one flamed out and had to be towed in from the dearming area at the runway's end. I wondered what the debriefing would be like.

It was short! Absolutely nothing was mentioned, except being weathered out of our primary and secondary targets. I kept my head down and my mouth shut for the time being. I was so wrung out from having to disobey a direct order and leave my flight that my blood pressure was probably sky high. I had never felt so angry and helpless in my life.

After the debriefing, I went straight to my squadron operations officer and told him to never schedule me to fly in the same formation with the higher-headquarters pilot again. When he asked me why, I said, "Because he is an incompetent idiot, and I will refuse to do so."

I never flew with the visitor again, and to the best of my knowledge, he never led another flight at Cam Ranh Bay.

Without a doubt, that was one of the most emotionally exhausting combat missions that I flew in Southeast Asia, and not a bomb was dropped nor a bullet fired.

I can't do this!

Central Highlands, RVN, 1966

As I remember, it was a divert mission from our prebriefed objective. Enroute to our target, the airborne command post had given us new target coordinates and a different forward air controller's callsign and radio frequency.

As we headed toward the latest coordinates, north of Pleiku, I was trying to guess what the new target would be and wondering if our ordnance load would be right for the tactical situation, whatever it might be. We were a flight of four Phantom IIs each with four one-hundred-and-fifty-gallon cans of napalm and a twenty-millimeter Gatling gun. Our napalm bombs were very effective for troops but not good for any sort of "hard" targets such as bridges, roads, and the like.

I had been in Southeast Asia quite long enough to feel pretty comfortable in my flight-leader role. Even as a relatively junior captain, I had been leading quite a few flights as of late. This seemed to happen more often on the early morning sorties than, for instance, the afternoon flights. I was not privy to the scheduling procedures, but I sometimes wondered if it was because I was a "teetotaler" and less prone to early morning headaches. Probably more likely it was the early wakeup calls. I also liked to imagine that it was my "outstanding" flying and leadership abilities that had shown through, but a more realistic view was probably the early wakeup option.

Whatever the reason, there I was leading four F-4s toward an unknown target situation with uneasy anticipation. I did know for sure that the folks with the "big picture" in the command post thought that we were sorely needed at the new coordinates. Whenever our targets were urgent enough to warrant a divert, it usually meant a tactical emergency for our ground troops, and that was a number-one priority with fighter jocks. Whenever we would receive information that another comrade was in trouble, it seemed to make the adrenalin kick in and the willingness to help become almost an obsession.

"Ground troops in trouble" came the urgent call from our forward air controller. A superior force of North Vietnamese regulars had infiltrated the area and were engaged with and about to overrun our badly outnumbered American troops. As we arrived in the area, we were told to standby while the forward air controller updated his information with the ground commander. While we orbited, I obtained a fuel and weapons ready check from my other three Phantoms. At the same time, I scanned the dense cover of trees below to try to ascertain some signs of the conflict. At times I thought that I could see dust and smoke from small arms fire and an occasional explosion from what was probably grenades. My heart was pounding as the adrenalin flowed freely, and my eagerness to help the Americans below just made it intensify.

The forward air controller suddenly broke the radio silence, "The ground commander advises that they are about to be overrun by the bad guys and he needs immediate napalm drops. The bomb passes are to be from the south toward the north. He will mark the area with purple smoke and wants the napalm to hit on a line between the two smoke locations, perpendicular to it, and all the way along it. The 'nape' needs to splash toward the trees just to the north of a line between the two purple smokes." I could see the purple begin to rise as the forward air controller called and said the ground commander had just reported the initiation of smoke.

As I positioned my Phantom for my first pass and called "Switches hot," I asked the forward air controller, "Where are our troops located?" I was concerned, of course, with hitting some of my own guys, which was a dread that we all lived with. The forward air controller replied, "The ground commander reports that his men are located along the line between the smokes." "But that's where you told me to drop!" I protested.

"That's right Whiskey lead, the commander on the ground wants the napalm right down on top of him and his troops. He says that they're dug in and the enemy is overrunning them. He says do it now or they're all dead anyway!" The feeling that came over me was like nothing I had ever felt before and like nothing I have felt since. I wanted to be anywhere else and doing anything else. I wanted to not be the flight leader and the first to drop. My mind told me it somehow wouldn't be so bad if I weren't first.

My F-4 was moving at five hundred knots as I rolled in toward the smoke and my mind was in turmoil. My hands were shaking like I had palsy, and I genuinely doubted that I could make my right thumb punch the bomb button. My gunsight moved rapidly across the ground toward the area between the slowly drifting smokes. I thought that I could see figures moving out of the trees and muzzle flashes accompanied by more small explosions. The gunsight was approaching the target. As it crossed the imaginary line, I blinked back my tears and punched the bomb release button. I felt the napalm release and my Phantom get lighter by about one thousand pounds of jellied gasoline.

"Good bomb, lead. Two Three and Four, put your bombs on both sides of Whiskey's lead," the forward air controller called. I tried to look back over my left shoulder as I made a hard climbing left turn, but my vision was blurred. I blinked rapidly to clear my eyes as my second napalm pass was rapidly approaching.

"Lead, ground says put your next one right on the east smoke . . . that some of the enemy are trying to make an end run," the FAC said. I suddenly felt a little better. At least the ground commander was still able to talk to the forward air controller. I lined up on the remaining wisps of purple smoke at the east end of the target area and obliterated it and everything else for hundreds of feet beyond it.

I don't remember anything else about the rest of that personally devastating mission. I didn't have a scratch from the incident, but somewhere within my soul I knew that I was deeply wounded with a scar that would always be there. I can still feel it as I sit here thirty years later and write this.

We received word later that day or the next that our mission had been a success. The North Vietnamese regulars had been caught mostly in the open, and their ranks were decimated by our napalm attack and by gunfire from the troops living through the "Hell" that I and my flight members had just created for them.

I don't know the answer to the question that's probably in your mind. I heard reports that there were only minor injuries suffered by our dug-in troops because of our attack. I sometimes suspect that the reports were fostered to protect our own psyche.

I heard later of other similar missions, and I know that there were countless other examples of such supreme courage by the ground troops in the field. I subsequently found out the identity of the Army captain commanding the men on the ground. I shall not use his name, but I salute him and his superb men, especially those who made the ultimate sacrifice for their comrades and country.

There is "no greater love."

Guardian angel?

Pleiku Air Base, RVN, 1966

The inspired pages of the *Holy Bible* tell about "ministering spirits." Many people believe this means each of us has a heavenly being assigned to us to watch over us or assist us in times of extreme trial or stress.

The weather over the primary target, a hidden truck park near the Ho Chi Minh Trail, was down in the weeds. In other words, the clouds were too low to find and strike the target. Our secondary target was farther from our home base of Cam Ranh Bay, RVN. It also was nearly hidden by weather, and by the time we located it and flew the airstrike, our fuel state was becoming critical. The leader of our flight of four Phantoms decided that a fuel stop at Pleiku Air Base in the central highlands was prudent, especially with the monsoon season upon us.

It was late afternoon as we climbed out of our cockpits and watched the refueling trucks hook up for our internal fuel loads. After a short briefing, we began preparing our fighters for engine start and departure. Pleiku had no ground-start units, so we were going to have to make cartridge starts. A ground-start unit is a small jet engine mounted on wheels. It connects to an aircraft's jet engine starter and directs air to the aircraft to spin the engine fast enough to start.

A starter cartridge accomplished the same feat by the slow, controlled burning of a special kind of gunpowder. It was like a huge shotgun shell. There was a starter breech under each engine in the F-4 where the cartridge was installed when needed. Each F-4 carried a starter cartridge in a space in each wing root. No problem, we made cartridge starts all the time when we were on alert. I set the cartridge for the left engine under the starter breech and went to the right side to remove that cartridge. I undid the fasteners and opened the storage area—you guessed it, no starter cartridge! Unlike some other aircraft that can crossbleed air from one engine to start the other, the F-4 needed a cartridge for each engine. I informed my flight leader of my problem, and we began asking the flightline personnel if another F-4 cartridge was on base—you guessed it again: No!

As the other three F-4 crews began strapping in their cockpits, I conferred again with my flight leader. He told me to go call the Command Post at Cam Ranh Bay and see what it could do. I told him I'd see him later and backed away while the F-4s started and taxied for takeoff.

I was on the phone to Cam Ranh Bay when my six squadron mates headed into the late afternoon skies toward home base. I envied them, especially when the command post told me they couldn't get a cartridge to me until tomorrow.

My backseater and I were really hungry. As we began chomping on a couple of burgers at the small officer's mess, I heard my name being paged. Picking up the phone, I heard the duty controller's voice from the Cam Ranh Bay command post. "Captain Cook, we've just been informed that Pleiku may be overrun tonight by Viet Cong sappers. The Commander wants you to get your F-4 out of Pleiku now."

"But I can't," I protested. "Does he know that I only have one starter cartridge?" I asked.

"He knows," came the reply. "The colonel said just tell him to get it out of there." As I hung up the phone in disbelief, after asking for and getting an authentication, the first trip-flare went off in the evening sky. Trip-flares are devices strategically placed around an area to assist in security. When a trip wire is touched by the enemy, a small rocket propels a flare into the sky. It then descends slowly under a parachute and gives off a tremendous amount of light. Next came the sound of small-arms fire. I hurried back to my table and broke the news to my backseater. His mouth dropped open and he said, "You've got to be kidding!" Another trip flare lit up the front of the room. He said, "Let's go!"

As we closed the starter breech on the left engine, the sun had just set behind the hills west of the air base. "You don't have to risk it with me," I repeated again. "I don't know if I can get this thing off the ground on one engine, if the other one doesn't airstart during the takeoff. It probably won't."

My plan was to hold the brakes with full afterburner power on the left engine and begin the roll with the flaps up to get all the thrust that I could. I would activate the air start ignition on the right engine with its throttle pushed up to military position. I hoped that the air passing through the engine as we rolled down the runway would be enough to allow an engine start and give us some thrust. About three thousand feet from the end, I would set one-half flaps.

As I rotated, I would hit the "panic" button that was designed to jettison all the external stores and bomb racks. I hoped that it would reduce the weight enough to let me fly. I gave my backseater one more chance to stay behind, but he wouldn't hear of it. I figured our chances were good that we'd have to eject at the far end of the runway, so I briefed him to be ready with his hands on the ejection handle. Hopefully, we'd have at least the minimum airspeed required for our parachutes to open.

More and more trip flares were going off in the dusk. Their smoke was drifting across the approach end of the runway as I prepared to start the left engine. Suddenly, out of the near darkness, came a camouflaged C-47 with absolutely no markings of any kind. It was banking sharply and in a very steep descent. In fact, it looked as if it were about to crash. At the last possible instant, the bank rolled out, the nose came up, and it made a beautiful landing.

The C-47 then turned off the runway and headed straight for us. As it passed in front of our F-4 and made a sharp left turn, the rear door opened. A figure wearing green camouflage fatigues with black Master Sergeant stripes jumped to the ground, looked at me, then turned back to the open door. He grabbed a cardboard box and walked up to the side of my aircraft. "Could you use this?" he asked and smiled. He then set the box down, walked back to the C-47, and disappeared inside.

The engines revved up, and the C-47 vanished back into the smoky sky as suddenly as it had come.

My backseater hopped down on the right wing and drop tank and hurriedly installed the starter cartridge. With both engines running, I called the tower for taxi clearance. The tower cleared me, and as I began to move, I asked, "Where did that C-47 come from?"

"We don't know," said the tower operator. "It made no radio calls and didn't get clearance to land or takeoff. We didn't even know it was around until it appeared over the end of the runway. Did you see any markings or identification numbers?"

"Nothing," I replied.

"What did they give you?" the tower continued.

"Probably my life," I thought, but answered, "A starter cartridge."

After I landed, I couldn't wait to get to the command post and thank those on duty. They were surprised to see me and assured me they had nothing to do with my "gift." I then queried all the other flight members and they had made no calls. In fact, they had assumed that the command post would handle it and had not given it another thought.

As I related my story to them, one of them laughed and said, "Maybe it was a ghost!" My mind went back to the biblical references about "ministering spirits."

After a short pause, I said, "I think maybe I know who it was, but I didn't know he wore Master Sergeant stripes."

Was it fate?

Cam Ranh Bay Air Base, RVN, 1966

Some people believe in fate, which can be defined as an unseen power that causes certain events to happen. In other words, things are predestined or programmed and cannot be changed.

I do not necessarily believe in fate, but some things that I have seen and heard have given me cause to wonder. Following are two examples.

It was a target in southern Laos, and it was "hot." Several aircraft had been hit in this particular area in the last two days. There was increased enemy activity on the Ho Chi Minh Trail, and it was obvious that the numerous antiaircraft guns that they had brought into the area were meant to protect that activity. Numerous trucks and other vehicles had been spotted moving along the trail toward this area during the last few nights, and it was obvious that something was up.

The area where the guns had been set up was known to be full of hidden truck parks and storage areas. The guns had first been reported by a forward air controller looking for trucks. One day they weren't there. The next day they were.

I don't know exactly how many guns there were purported to be, but they were numerous. The target briefing said that they weren't even very well camouflaged. That probably indicated they felt with that many guns, they could protect each other with the cross coverage. It was also an indication that they probably had hauled plenty of ammunition down the trail to keep the guns firing. We were scheduled as part of a series of airstrikes on that particular day to try to "take them out."

I was leading the flight of four Phantoms. It was a beautiful sunny day all the way to southern Laos. The target area was clear also. It was easy to find the guns. As soon as we arrived in the area and began circling to locate them, they obliged us and started firing. I was really surprised at their number. There were many muzzle flashes from several locations. They certainly didn't appear to be shy about revealing their location.

Our weapons of the day were seven-hundred-and-fifty-pound dive bombs and the twenty-millimeter Gatling gun. Using the gun against these guys wasn't even a consideration to me. I had been taught to never intentionally get into a gun duel with numerous anti-aircraft guns unless you have a "death wish." I didn't, so it was going to be strictly dive bombs. There didn't appear to be much wind. That made our bombing computations easier.

After the usual fuel-and-weapons check, I rolled in on my selected target. It was easy to choose. I just used the heaviest concentration of big muzzle flashes for my aim point. My backseater called off the altitudes as we dived through them and I "pickled" off my first 750-pounder when he gave the word. There were shells going off everywhere around us as we came "down the chute," so I was doing some high-G "jinking" (hard turns to spoil the gunner's aim) as we came off the bomb pass. I looked down and saw my bomb hit just to the right side of my intended target. I had undercorrected slightly for the estimated wind. I called that information to the rest of my flight so the pilots could adjust their aim points accordingly. Fortunately, the seven-hundred-and-fifty-pound bombs were large enough to do substantial damage to a target even without a direct hit. (A hit within one hundred and forty feet of your intended target was considered a qualifying bomb on a stateside gunnery range, but, of course, the closer the better.) I would adjust my next aim point a few meters more upwind.

My Number Three F-4 reported a hit on his first pass. The pilots were both okay, but they had lost one of their generators and were having some minor electrical problems. I immediately told him to head for home or to go to Da Nang if he needed to. Da Nang was almost due east of us and a lot closer than Cam Ranh Bay. I sent Number Four with him to assist, if needed.

I rolled in on my second pass as I ended my radio conversation with Number Three. A few rounds whizzed past the cockpit just before bomb release. These guys were good. I got a better bomb this time and definitely silenced one gun site. We got through "clean" (without being hit) again and so did Number Two.

As I was on an eight-thousand-foot downwind trying to select my next target, I got a call.

"Hey, two Air Force F-4s just pulling off the guns at (name of area), this is Navy A-4 (callsign).

"Go ahead Navy, this is Whiskey flight," I answered.

"Hey, I like that callsign, Whiskey! We're two A-4s. We've each got a 'Bullpup' hanging. We're approaching bingo fuel and wonder if you'd be willing to let us have a go at these 'Gomers' while you sit back and relax a spell," said the leader of the little Skyhawks.

I couldn't think of anything I would have liked better at the moment, and I relayed that to the "Navy jocks." I also told them that I enjoyed nothing better than having someone do my work for me, especially a couple of Naval Aviators.

I pulled my flight of two off high and dry (and out of range of the guns).

Bullpup missiles were fun to launch and entertaining to watch. They had a small conventional warhead. They were launched from a fighter and then steered to the target with a little control stick in the cockpit of the launching jet. There was a flare on the rear of the missile to assist the pilot in keeping it in sight as he guided it to the intended destination. Two problems with them were the relatively small size of the warhead and the fact that the launching aircraft had to fly down a near constant flight path toward the selected target. This was because the guiding pilot had to keep the missile in sight at all times during its flight to the target. He had to make a myriad of corrections to the missile's guidance, the main one to overcome the force of gravity.

The constant glidepath by the launching aircraft was no problem on an undefended target. The target today was not undefended.

The lead A-4 pilot launched his Bullpup and followed it down toward one of the firing gun positions. There were antiaircraft rounds going off all around him, but he held his Skyhawk steady and guided his "Pup" right into the middle of one of the gun positions. That particular gun position "disappeared." He pulled off unscathed.

Number two was coming down the chute following his Bullpup toward its target. I was looking at the missile from my advantageous position when I saw the flash behind it. I quickly looked at the little jet. The A-4 was losing fuel out of its belly. The fire had extinguished immediately after the initial flash. The pilot had no choice. He was being surrounded by antiaircraft fire. He aborted his pass and pulled off, turning hard to evade the guns which were all trained on him and trying to finish him off.

He managed to miss all of the rounds flying at him and escaped toward the east. His leader caught up with him and looked him over. He was still losing jet fuel at an enormous rate. They headed for Da Nang Air Base on the eastern coast.

We were approaching our "bingo" fuel (the predetermined fuel-remaining when an aircraft has to start toward its planned landing field). I called my wingman and told him that we were going to make only one more pass and to set his bomb switches to drop all of his remaining bombs.

The gunners had been encouraged by their hit on the Navy attack fighter and let go with everything they had.

They missed. Both my wingman and I came through the last pass "clean" and hightailed it thankfully toward home.

I wondered if my number three F-4 had made it home. (The number three F-4 of my flight and his wingman had landed safely at Cam Ranh Bay with no further problems)

I wondered if the A-4 pilot had made it to Da Nang. He had run out of fuel about halfway. He ejected successfully while his leader watched and began coordinating a rescue. All that was available in the immediate area was a nonrescue helicopter (no rescue harness and cable winch). Because of the terrain, the chopper could not land. The chopper crew threw the downed A-4 pilot a rope. He grabbed it and hung on as long as he could. It wasn't long enough. Fate?

Fate two? They were from one of our sister squadrons at Cam Ranh Bay. They had been shot down and had ejected successfully from their F-4 just the day before. After their return to home base, the flight surgeons had checked them over and declared them fit.

Their squadron commander decided that the two needed a little break from the action and told them to get out of the country for a few days on some "R and R" leave. The required paperwork was signed and ready. They were elated at the news and wasted no time. They contacted the flight operations folks and found that there was a Marine C-130 "Hercules" landing shortly and then heading "out of country." There were seats available. They grabbed some civilian clothes, threw them in a bag, and headed for the flightline.

They were just in time. The C-130 was on the ground and getting ready to takeoff for Taiwan. They scrambled aboard and sat in a couple of the web seats located along the sides of the cargo aircraft. They couldn't believe their luck at hopping a ride so quickly. This would save them a full day of "R and R." One more day out of Vietnam.

The C-130 blew up just after takeoff. There were no survivors. Fate? You tell me.

Quit leaving this ___ on my runway!

Cam Ranh Bay Air Base, RVN, 1966

The trailer was hot and the magazines were six months old. The boredom was more oppressive than the heat. The fire-resistant "nomex" flightsuits were sticking to us like plastic wrap.

Outside, the four F-4s sat hunched over their bomb loads like mother hawks guarding their young. They were mean-looking machines and as capable of destruction as they appeared. The heat waves rising from the AM2 aluminum mat made them appear to move, to shift back and forth like something alive.

They were out there, and we were in here for one purpose, rescue. Rescue! With five-hundred-pound dive bombs and napalm and twenty-millimeter Gatling guns? The 12th Tactical Fighter Wing kept them on alert twenty-four hours a day. Two were assigned a five-minute scramble time with two more on fifteen-minute backup. The Army in the field was in constant danger of being overrun or pinned down by sometimes superior numbers of the Viet Cong, or, on occasion, by North Vietnamese regulars. When this happened, they needed help and they needed it immediately!

I was half dozing when the hotline rang, "Scramble two, vector three four zero!" the alert monitor yelled. Six of us hit the ramp at approximately the same time. Six starter cartridges fired and spun six big J79 jet engines less than a minute later. My crew chief visually checked my ejection seat pin out and gave me a thumbs up as he ran for safety through the cartridge smoke and I shoved the throttles forward.

"Canopy down and locked, light out," called my backseater as I did the same. Whiskey 02 was moving to my right, and 03 was waiting in place to see if either of us aborted. At one hundred and sixty-five knots, I pulled the control stick back and raised the landing gear handle passing one hundred and ninety knots.

"Two's airborne," Wayne called. I "Rogered" and began a left turn to heading three four zero degrees. Two hundred and fifty knots—the

Phantom seemed to stop as I pulled the engines out of afterburner. When it's hot enough and it's heavily loaded, even an F-4 is a dog sometimes. Three hundred knots—Wayne joined on my right wing. I looked over, and he gave me a thumbs-up, indicating all my ordnance was still attached. He then banked sharply right while I checked him. His napalm cans and gun were still intact, and I indicated so with a hand signal.

"Three four zero degrees, six zero miles, Sidewinder," were the final instructions from the control tower. We switched to the forward air controller's frequency and checked in. Sidewinder was the collective callsign of the forward air controllers in the area where we were headed. They flew little Cessna 0-1 Birddogs. Their courage while flying those little aircraft in harm's way was phenomenal.

"Sidewinder 25, this is Whiskey 01 with 2 birds, 'slick' 500 pounders, 'nape,' and 'pistols,'" I called.

"Roger Whiskey, Sidewinder 25, we have a downed chopper with seven injured. They are pinned down by an unknown number of bad guys. The chopper is on its side at the southwest base of a small hill. The Charleys are located in numerous trenches on the hill, just 30 meters east-northeast of the chopper. We're receiving heavy ground fire of various types. Say your position."

"We're ten miles southeast of you descending through fifteen thousand feet. Say the weather, we're above an overcast," I answered.

"Roger Whiskey 01, we have approximately a three-thousand-foot overcast, visibility one-five miles, wind south-southeast at about ten knots. There are friendlies on all sides, except the north. Suggest you run in south to north," he replied. "Which bird has the dive bombs?" he asked.

"Roger, 01 has them. I'm not going to risk dive bombs with the chopper so close to the target," I said.

I then called my wingman. "Whiskey Two, I'll lead you on each pass with the pistol. Maybe that will keep their heads down while you make your runs."

"Roger lead," Wayne answered. "Pistol" was the nickname for the twenty-millimeter Gatling gun we carried underneath the belly of our Phantoms. It fired at the rate of one hundred rounds per second and was a weapon feared by both the Viet Cong and North Vietnamese army.

"Sidewinder 25, I'm orbiting at twenty-five hundred feet to the north. Where are you from the target?" I asked.

"25 is three clicks west at fifteen hundred feet," he answered (a click is a nautical mile). I looked west and saw the little 0-1 as it banked back toward the target area. "I've got you in sight, 25," I called.

"Roger Whiskey, I'll mark the target now. They've stopped shooting since they heard you. They're probably hoping that you can't find them." I watched as the 2.75-inch rocket left the little Cessna's wing and plowed into the side of a small hill.

"Got the smoke, 25," I said.

"Rog, the smoke is ten meters west of your main target and twenty meters east of the chopper. Do you see the chopper?" he asked.

"I've got the chopper. Where are the good guys?" I said.

He answered, "About 40 meters west of the chopper, slightly down the hill. They've tried to put the chopper between them and the Charleys."

"Two, do you see the chopper?" I asked.

"Got it, lead," two answered immediately.

"Whiskey 01, you're cleared in 'hot.' They've started shooting again since I marked them. I'm ziggying to the west about three clicks, so don't run over me," Sidewinder called. "I'll orbit there during your passes."

"Wouldn't hurt a fly," I answered as I rolled in and called "Gun switches hot." I began to see the lines of trenches zig-zagging across the hill. I pointed the nose of my Phantom toward them. The bad guys decided then that I had them in sight and the muzzle flashes covered the hilltop. Most of it looked like thirty caliber, but there were a few larger flashes interspersed also. I squeezed the trigger and the first trench disappeared in a cloud of dirt and smoke as the high-incendiary rounds found their way to the ground. I released the trigger after three seconds and reefed the control stick back and left, as I started my jinking maneuver. I felt the airframe shudder and two or three metallic clunks as several rounds struck my F-4. I jinked back right and could see Wayne's Phantom's belly as he jinked right also. Behind him, part of the hilltop was on fire.

"Nice work, 01, put the next one twenty meters closer to the chopper," Sidewinder called.

"Copy Two?" I asked Wayne.

"You got it, Sidewinder," he answered.

I called "Leads in," as I pulled the big fighter into a five-G vertical bank at five hundred feet. "Are you still with me Two?" I asked.

Wayne grunted under the G forces, "I'm tight at your six lead. Show me the target!"

I looked through the gunsight and the hillside lit up again with ground fire. I moved my aim point more toward the chopper and squeezed the trigger.

"I sure don't like running in the same direction again," my backseater said.

"Neither do I, but we have no choice," I answered. The brown earth swept underneath the charging Phantom's nose as I pulled back and left into an eighty-five degree banked left jink. The airspeed indicator was showing four hundred and seventy knots and increasing as I reefed into a hard right climbing turn at full military power. No metallic thunks this time as we came through clean. I glanced up in the mirrors, and Wayne's F-4 almost filled them as he closed the gap between us, well out of range of the ground fire.

"Whiskey 01, suggest you make this your last pass. We have some Sandys and a Jolly Green just south, waiting to pick up the chopper guys. Looks like one more sweep like the last one might be enough to get them out," Sidewinder 25 said. (Sandys were the propeller-driven A-1 fighters who flew armed escort for the Jolly Green rescue helicopters.)

"Rog, Sidewinder, we'll make this our last," I answered.

"Two copies," Wayne acknowledged. "I'll salvo my remaining cans."

"Put the stuff right between the first two runs, Whiskey," Sidewinder 25 said. "The chopper guys say that should do it."

"You got it Two Five," I answered. The wingtips on my Phantom shuttered in protest as I squared off the corner slightly inverted and lined up the gunsight. I looked back and could see the vapor trails off the top of Wayne's Phantom as he dug in for the last run behind me. Through the center windscreen, the hill loomed larger and larger as I waited for the proper slant range to fire. I stabilized the airspeed at four hundred and sixty knots and squeezed the trigger as the gunsight centered on the heaviest gunfire. The hillside once again exploded in flashes and dust as the twenty-millimeter found its mark. I pulled back on the controls, stopping my rapid descent.

Instantaneously, the world tilted violently one hundred and ten degrees to the left. I reacted with full left stick and shoved it forward to push the hilltop, fifty feet from my canopy, farther away from my face. I had stopped the rolling motion at about one hundred and twenty degrees of bank, but now we were slightly inverted at around fifty feet in the air. I began running the stabilizer trim toward the nose-down position. That helped my Phantom begin an inverted climb away from the ground, now just a few yards away.

Both myself and my backseat pilot were hanging in our shoulder harnesses. The top of the canopy was filled with sand and dirt that had accumulated in the floor. I could not reach the rudder pedals, and my full left and forward control stick inputs just served to keep us in the same semi-inverted position. As the F-4 exchanged its airspeed for more altitude, I finally got the toe of my left boot against the rudder pedal and pushed, straining to make my leg longer. The rudder finally began to work against the airstream and we began a slow, yawing roll to the left, until the wings were level again.

I first checked the engine instruments, and they all seemed normal. I then looked at my right wing and saw the cause of my near demise. Ground fire had exploded my right external fuel drop tank, which normally hung suspended under the wing. Now, however, it was on top of the wing, and opened up like a big metal fan. The violent disruption of airflow had caused my right wing to lose a tremendous amount of lift, simultaneously causing a large yaw to the right. This then increased the effective lift on my left wing. The result was an instant, uncontrolled roll to the right at just fifty feet of altitude. If we had been hit a fraction of a second sooner, our descent would have slammed us into the hilltop.

Wayne's voice in my headset was urgent. "Whiskey 01, are you guys okay?"

"Yeah, except for some sand in our eyes," I answered.

Wayne had joined on our right wing and began helping us assess the damage. In addition to the drastically modified fuel tank, we had two large holes blown out of our right-wing leading-edge slats and numerous holes underneath the right wing. Also, Wayne said it looked like the paint on the right side of the vertical tail was singed.

(He subsequently related that he had seen the right side of our F-4 engulfed in a fireball and the violent right roll. He thought we had hit the ground because of the dust cloud we had kicked up. But then he saw us emerge from the cloud and fire in a near-inverted climb and had been amazed that we were still flying.)

My next problem was going to take longer to analyze. Could I control the aircraft for landing or would we have to eject? Slowing to two hundred and fifty knots, I turned the Phantom toward home and out to sea. If we had to eject, I wanted to be over water. Water was the only sure safe area in that part of Southeast Asia, if we got far enough away from shore and the numerous small boats.

Enroute to home base, I began to experiment with my fighter. I discovered that with full-left aileron trim, I could keep the wings level down to about two hundred knots. That was sixty-eight knots above the one hundred and thirty two that was our normal landing speed. I elected not to try to lower the slats and flaps. I was concerned that attempting to lower the damaged slats would rupture my hydraulic system. Besides, part of the right fuel droptank was still wrapped around the leading edge of the wing, and I wasn't sure that the slats would lower anyway.

With my emergency condition declared, the control tower cleared me to land. With full-left rudder trim and near full-left stick, the big F-4 felt very strange and unstable. Also, the runway was really rushing at me with my two-hundred-and-ten-knot approach speed. I wondered if I would be able to get stopped okay. In addition, I hoped my right landing-gear tire was inflated as Wayne had told me it appeared. We could still end up a statistic with a flat tire at two hundred knots.

The touchdown was one of my smoother ones, and I pulled the drag chute handle at two hundred knots, its maximum-design speed. The tug of the deploying chute was welcomed and seemed much more effective at the higher-than-usual speed. I gingerly touched the brakes at one hundred knots, and everything felt normal. The Phantom slowed rapidly, and I was relieved that I wouldn't have to engage the emergency barrier with the tailhook.

As we sat in the cockpit in the dearming area, I looked back at the runway. There was a blue staff car on the runway, stopping every now and then. Someone was getting out of the driver's seat and appeared to be inspecting the surface of the runway or perhaps picking something up. The staff car with the colonel's eagle insignia on the front slowly pulled up beside my aircraft. Our Commander stepped out, glared up at me, and opened the rear door. He reached inside and turned around with an armload of F-4 parts that my Phantom had evidently shed as it rolled down the runway. The colonel looked up at me and yelled. Although the engine noise was drowning out most of it, I could read his lips: "Cook, quit leaving this s___ on my runway!" With that, he turned around, threw the scrap metal in the backseat, and slammed the door. Turning toward me again, he grinned, waved, got in his car, and drove away.

Another day, another way

Coastal Lowlands, RVN, 1966

Some of my memories from Vietnam are hazy, almost like a television picture slightly out of focus. Others stand out so sharp and clear that it seems the incident could have happened yesterday. One of the vivid ones is especially clear. I can recall it with amazing clarity whenever I choose.

It was another day mostly spent trying to find something to pass the time in the alert building near the south end of the runway. The big Chrysler window air conditioning units were in high gear trying to keep a survivable temperature inside the underinsulated port-a-building. Most of the guys had their flightsuit sleeves shoved up and the zipper tab pulled down to their waists trying to stay as comfortable as possible.

Others, including myself, had completely pulled down the flightsuit top and had the sleeves either hanging down or tied loosely around our waists. I had arrived in Vietnam weighing two hundred and seven pounds, but my excess "table muscle" was rapidly departing. I was down to about one hundred and ninety already. (I weighed one hundred and sixty-seven pounds when I went home in October, but I wouldn't recommend it as a good way to shed those unwanted pounds.)

I read all the same outdated magazines the last time I pulled alert. I believe in the power of positive thinking, and I was positively thinking that I wished the scramble phone would ring. That way I could go fly a combat mission and momentarily escape these surroundings before the heat and boredom killed me. It rang.

"Scramble two, three two zero, eighty." I was out of there. As I ran across the hot aluminum ramp toward my waiting Phantom, I was twisting and turning my upper body trying to pull my flightsuit sleeves onto my sweaty arms. I remember a pencil or pen falling out of a sleeve pocket and almost being trampled by the crew chief running behind me as I stopped to retrieve it. It kind of reminded me of an old Keystone Cops movie for a second or two. The deft crew chief avoided the collision by a fraction as he flew by me. He was already pulling the bomb arming pins as I scrambled up the ladder on the left side of the Phantom's front fuselage. As I settled into the front seat, he was up the ladder and "last checking" my fittings and safety pins. I saw his thumbs-up signal as he scrambled down and ran to escape the acrid smoke from the starter cartridges as I fired both engines simultaneously.

We made the five-minute scramble-time window with just a few seconds to spare. Everything seemed to move so much slower when it was this hot. That included an F-4. I've always thought that it is an interesting phenomenon that many machines are affected by heat and altitude in much the same way as humans. Even as powerful as the engines were on an F-4, they were effectively "derated" by the heat. I don't know what the actual numbers would have been, but I guarantee that the big General Electric engines were not producing thirty-four-thousand pounds of thrust on a hot day like that one.

"Three two zero, eighty," the tower repeated our scramble instructions, which meant "heading three hundred and twenty degrees for eighty miles." He also gave a forward air controller's callsign and a radio frequency to contact him on. As I always did, I speculated as to what our target would be and if we would be carrying the right weapons to do the job. It felt good to be in the familiar cockpit with the air blowing cooler now that we were at altitude. I had once again escaped death by boredom.

In just a little over ten minutes, we were circling the target area and talking to the forward air controller about our target. It was a cluster of several buildings located on the west side of a north-south treeline. To the east of the trees was a large rice paddy, and some scattered trees were on its eastern border. An Army patrol had come upon a large number of Viet Cong. They were outnumbered by the Cong, so they had wisely chosen not to engage them; however, they had been able to follow and observe them enter the cluster of buildings. They were our target. We were to be their airborne artillery.

The forward air controller decided that our twenty-millimeter Gatling guns were the weapons of choice. Between our two Phantoms, we carried about two thousand rounds of explosive shells. The firepower from our guns firing at the rate of one hundred rounds per second was devastating. The bullets leaving the six rotating barrels actually produced almost four thousand pounds of negative thrust. It was enough to cause a slight dipping of the Phantom's nose. We actually liked that particular effect because it helped to keep the gunsight on the target as we moved across the ground toward it. The impact of the bullets hitting the target was truly fearsome. I heard stories that captured enemy troops stated that our Gatling guns were their most feared weapon. I believe their statements were something like "Light in sky, everybody die!" The "light" of course was the constant muzzle flash as the gun was fired.

After two passes, the immediate target area was almost obliterated. As I was rolling in for my third run, the forward air controller told me that several Viet Cong soldiers had escaped into the treeline to the east side of the buildings. He said to change the target to the treeline if I had time to correct my run-in. I did. I pulled the nose hard right about fifteen degrees and rolled out just as I got in range to fire. I lit up the treeline for about one hundred yards with a three-to-four-second burst. I could see a few muzzle flashes from ground fire but nothing substantial.

My flight leader had "fired out" on his last pass and had pulled off "high and dry," out of ammunition and orbiting the target area out of the way. The forward air controller asked me if I had any ammo left. I told him I should have two to three hundred rounds left. He said "Good, one of the Cong has emerged from the southern end of the trees and is running east across the rice paddy." I thought to myself that he would have been much better off and smarter to have stayed in the relative cover of the trees. Now he was in the open, and the Gatling gun was very accurate even when we were moving at nearly five hundred knots of indicated airspeed. As I had expected, the forward air controller said that the running figure was my last target.

I remember thinking that I was carrying an awful lot of firepower for one lone soldier and wondered why the ground forces didn't just go after him. I suspected that they were out of position or might be afraid of ambush. I also knew that the forward air controllers took a lot of hits from the Cong and suffered constant casualties from their ground fire. I didn't blame them for wanting to get rid of all the Viet Cong that they could. I also knew that the forward air controllers were familiar with their areas and the local military situation. They were the "boss" at such times. That's why they were called "controllers." It didn't matter what I thought.

I rolled in. I could see the burning buildings and several trees on fire. I strained to see the running figure in the rice paddy. I finally spotted him in a loping, bent-over run toward the trees to his east. Something was really bothering me about the whole situation. Was it just the basic unfairness of one lone soldier in a rice paddy running for his life from certain total annihilation by an airborne six-hundred-mile-per-hour twenty-millimeter cannon? Certainly that was part of it. But there was something else, something in the way he was moving. Something in the way he was bent over. There wasn't much time. I was almost in firing range with the "target" in my sights and my finger on the trigger. It *was* time.

I jerked my finger off the trigger and racked the Phantom hard right, then even harder left. I was very low. My left wing tip was almost in the rice paddy mud as I swept by the now motionless figure. We looked at each other as I roared by. I was having to peer through the top of the canopy because of my almost ninety-degree bank, but my mind clearly snapped the picture that I can still recall at will.

He must have been eighty or ninety years old. His white beard was almost to his waist. I silently thanked God and pulled into a steep climb and called "Two's off, dry."

As I stared back down over my left shoulder, I could see him still standing there. He had raised one hand. I don't know if he was shielding his eyes from the sun or if it was some kind of gesture.

That was thirty years ago. That old gentleman has to be in a grave somewhere in South Vietnam by now. I didn't put him there. I can still "see" him anytime I want. It's always a blessing.

A mind is a terrible thing!

Clark Air Base, Philippines, 1966

Of course, the rest of the well-known phrase heading this story is "TO WASTE." This chronicle is about a "stunt" that I pulled at Clark Air Force Base in the Philippines while on Temporary Duty from my permanent base at Cam Ranh Bay, South Vietnam. My backseater and I were attached to a maintenance section at Clark Air Base whose job was to repair battle-damaged F-4s out of Vietnam. Our job was to test fly the repaired Phantoms and ascertain whether they could be returned to service.

The commanders at Cam Ranh Bay were using these short TDYs to get their aircrews out of combat flying for a while. I think it was only a week or so that we spent in the Philippines each time.

One evening after having dinner at the Clark Officer's Base Club restaurant, we went downstairs to the bar. That was the congregating place for all the aircrews. We were sitting at a table when my backseater heard his name being called. He recognized the "hailer" as a guy he'd gone to pilot training with. He decided to go over to say hello. His friend, a bachelor, was sitting at a table with a female schoolteacher from the base school for the permanently assigned personnel's children. There was another pilot standing at the table in his flight suit. He had a drink in his hand and was also talking to the schoolteacher.

After my backseater had stood there for a couple of minutes, he motioned me to come over. As I approached, I noticed that the pilot who had been talking to the schoolteacher looked at me with a kind of cocky sneer on his face. As I got to the table, I saw the organizational patches on his flightsuit. He was an F-100 Super Sabre "jock."

Now it must be said here that there was a usually friendly rivalry between the Super Sabre pilots and the F-4 Phantom "Phlyers" as we were sometimes called. As my backseater finished introducing me to his former pilot training acquaintance, the F-100 pilot said, "So you're one of those damned F-4 drivers."

I said something to the effect of, "Well I've been called worse by better."

He sneered again and said "How do you like driving around in a two-man bomber?" One must understand that was the main point of pride for all F-100 pilots. They were flying a single-seat fighter. I'll admit now, as would most "real" fighter pilots I know, that I'd rather fly in a single-seat fighter myself. There's just something about being in the jet by yourself. Maybe it's because you don't have to worry about anyone else in your airplane. That's the noble sounding reason. More than likely it's because there's no one there to see you if you "Let it hang out and then step on it!" (All fighter pilots and most men will know what I mean by that.)

Of course I wasn't going to admit to him what I just admitted to you. Instead I said, "It must be terrible to be flying a half-assed, excuse me, a half-'fast' over-the-hill limited-to-four-Gs collection of worn-out aluminum. (The F-100 would do about 850 knots on a real good day. It had been limited to four Gs for a while after a pilot "pulled" the wings off of one while on a mission in South Vietnam.)

He leaned toward me and raised his voice. His face was now a bright red and his blond crew cut seemed to stand up straighter as he said, "I don't have anyone in a back seat to push me faster!"

I leaned forward myself and stated, "Friend, you don't have *anything* in the back pushing you!" (I was referring to his engine, of course, which is located in the rear of the jet.)

My backseater, obviously more grown-up than I at the moment, basically demonstrated an earthbound and reversed version of what the F-100 pilot had just stated. He stepped between us and gave me a slight "push," but backward instead of forward. That served to open the rapidly closing space between myself and the other fighter pilot somewhat and effectively served to defuse the volatile "tactical" situation. (Remember, I already told you fighter pilots ain't grown men, sometimes.)

We left the club and went to the Visiting Officer's Quarters (VOQ) and hit the sack. We had an early test flight the next morning.

After breakfast, we reported for our briefing and flight. After a thorough walkaround preflight, we decided that the Phantom looked flyable although it had obviously been "rode hard and put up wet" a few times. There were big patches of heavy aluminum on the trailing edges of the wings just inboard of the wing-fold areas. There were other various patches and scratches around, but nothing was leaking out on the ground or missing, so we went flying.

Another reason for wanting to come to Clark and fly test hops—besides getting out of "Nam" *and* arguing with Super Sabre jocks—was the "clean wings." We carried no external fuel tanks, no bomb racks, and no missiles. We got to fly an F-4 "lite" you might say. It was awesome! It was a completely different jet when it wasn't loaded down. (You might have heard that flying a hot jet fighter is better than sex. Let me correct that observation. That statement was obviously made by someone who hasn't tried either one or the other. But it *is* a close second.)

I released the brakes and jammed the throttles into full afterburners. The Phantom leaped forward like something wild. It had been a long time since I had gotten to fly a clean machine. I had forgotten how that big ugly sucker would go. I thought of my intelligent "discussion" of the previous evening as I pulled back on the stick. We were already hauling at 300 knots. I thought "I'll show that smart ass" and hoped he was either watching or that I was waking him up with my afterburners. We rolled upright from an inverted level-off at around thirty thousand feet and began accelerating. I slipped the throttles out of afterburner.

Quickly we ran through the test-flight checklist, and the F-4 had passed with "flying" colors. Although she was looking a little worn on the outside, she was sweet as pie on the inside. (Suppose there's a moral there?) I lit the afterburners again, pushed the nose over, and put the jet into a one-half-G condition. It was at one-half G that the Phantom II was found to accelerate the quickest. I asked my trusty backseater what was the highest altitude he'd ever been. I don't remember his answer, but I think it was about the same as myself, which was somewhere in the high forty thousands.

Our speed was approaching around Mach 1.9. I hadn't been that fast since I had checked out in the bird. I started easing the stick back gently so I wouldn't "bleed off" any unnecessary speed. I kept pulling it back until the Phantom's nose was about sixty or seventy degrees above the horizon. We were going up like a rocket at first, but as we entered the sixty thousands, the airspeed was dropping rapidly. At around sixty-five thousand, the afterburners started pulsating. I gently pulled the throttles out of the afterburner range and started slowly pushing the nose over.

Then it hit me! Remember the name of this particular story? Here we were sitting in a battle-damage-repaired F-4 just coming over the top of a zoom climb at nearly seventy-one thousand feet with no pressure suits! If the engines had flamed out at that altitude or even a lot lower, we were dead! If you lose your cabin pressure at those altitudes, your blood's makeup drastically changes, and you're in serious, serious, circumstances.

I couldn't wait to get down. I know I was sucking up oxygen at a tremendous rate, and I could hear my backseater over our hot microphone doing the same. I wondered if he was thinking that I was as stupid as I was thinking that I was! I had the nose pointed straight down out of the clear blue sky and at the lush green landscape of the Philippines. We were coming down a hell of a lot faster than we went up. As we passed around fifty thousand feet, I began to breathe a lot easier. I began to silently revel in having proved the old adage that, "Once again ignorance and stupidity had won out over skill and science."

"A mind *is* a terrible thing if you don't use it!"

I claim it was combat fatigue. At least that's my story, and I'm sticking to it!

Although I dearly love the F-4 and what it could and did do, I confessed earlier that I would prefer flying a single-seat fighter (especially if it performed like a Phantom II). What do you bet that my backseater wished? I know *he* would have rather been flying one!

Postscript. That same day after landing my "high-flying" F-4, I went to the Officer's Club to get some lunch. There were several people enjoying the bright Philippine sun around the pool. There were some little kids splashing around with their moms and having a ball. I thought about my son Jeff and my as-yet-unborn child back in Florida.

I had walked up to the edge of the pool and was still deep in thought when someone called my name. I looked up, and a guy was waving at me. He was in the process of standing up from dangling his legs in the water. He had been sitting on the other side of the pool from me, and I squinted against the sun reflection to identify him.

The shock that hit me felt like a physical blow. I know my mouth dropped open as he walked around the edge of the pool toward me with his hand extended. He walked right up to me and grinned. When I said nothing, he frowned.

"What's the matter, Jerry? You look like you've seen a ghost," he said.

His statement couldn't have been more appropriate. That is exactly what I thought I was seeing. I had been told just two weeks prior that this pilot standing before me had been shot down in his C-123 and killed.

"I thought you were dead," is all that would come out of my mouth. He grabbed his wrist with his fingers, checked his pulse, and happily announced, "Nope. I'm alive!"

We had a lighthearted lunch together celebrating his nondemise—he was relieved too!—and catching up on old times. We had been instructor pilots together at Laredo AFB for several years. I had lost track of him after leaving Texas for F-4s at MacDill AFB. The first that I had heard of him was the death report from a mutual friend.

It was obviously mistaken identity. I would be happy to relay that to my informant. I knew that he would be as relieved as I. But, *someone* had died. I was ashamed at not thinking that thought sooner.

Get it out, get it out!

Cam Ranh Bay Air Base, RVN, 1966

The first six months or so of our fighter wing's tour in Southeast Asia was almost exclusively absorbed by our air-to-ground mission. One of the major reasons for this was our location. We were approximately halfway down the coast between the Demilitarized Zone and the southern tip of South Vietnam. The DMZ was about a three-hundred-and-fifty-mile flight north of us. With two three-hundred-and-seventy-gallon external fuel tanks, bombs on the TERs (triple ejector racks), and a centerline gun pod, we could reach just into North Vietnam and the southern part of Laos (of course officially we weren't going to Laos at the time). Near the northern edges of our nonrefueled combat radius we sometimes had just enough fuel to "make one pass and haul ass." The point being we were restricted in our possible missions by the lack of air refueling tankers in our area.

The word came down that the situation was about to change. I don't know where the tanker resources transferred from, but we were going to finally have some air-to-air refueling availability.

I was really elated at the news, especially when they told us to start honing up our air-to-air skills when possible because we were going to start getting some MiG CAP and escort missions into the North Vietnam "arena."

Immediately, most flight leaders began trying to arrive back in the Cam Ranh area with enough fuel remaining to practice some "air-to-air." We would head out to the east of the base over the South China Sea and set up some intercepts and then some "turning and burning." That meant closer in maneuvering to try to get into position for a Sidewinder missile shot.

The Aim-9 Sidewinder used infrared tracking to guide to an enemy aircraft's main heat source, the engine exhaust. The Sidewinders of that era had an effective range of around two miles and a speed of over Mach two. The more you could place your aircraft directly behind an adversary, the more accurate the missile and the better the odds of a hit.

Wayne and I returned from a strike mission with enough fuel for a short "furball" (dogfight). We separated a few miles and turned in to try to maneuver behind each other.

We met in a head-on pass and each of us pulled the noses of our Phantoms up to shorten our turn radius. We were both in full afterburner to try to maintain our airspeed near the optimum for turning the Phantom. As we grunted and pulled as many Gs as the F-4's dynamic energy would allow, we ended up almost directly across an imaginary circle from each other. As our speed deteriorated we would lower the nose a little to try to maintain it. When the Phantom regained a little energy surplus, we would pull the nose up to try to shorten the turn radius. Neither of us was making much headway.

(Fighting another pilot who is flying the same type of equipment as you takes away a large part of air-combat training—that being the ability to utilize your aircraft's design advantages over your adversary *and* attempting to keep him from using his jet's advantages over yours. We were not allowed to air-combat train against dissimilar aircraft in those days. They were concerned that we would lose some aircraft practicing against other types. I guess MiGs didn't count.)

I was "pulling" the maximum performance according to my angle-of-attack indicator. I was in full afterburner. The jet was turning its best with the wingtip shudder feeling just right. The speed was right. The air was smooth. I was sweating and grunting with the Gs and having a ball. My "fangs" were showing. Life was good!

The nose of the jet pitched upward, and it rolled up and over to the right. The South China Sea and the sky swapped places a few times, and then the Phantom settled into a slightly nose-low rapid rotation to the left. I pulled the throttles to idle power and began the spin recovery procedures. We were about twenty thousand feet above the water.

"Get it out, get it out," came the urgent call from Wayne's Phantom as he circled my spinning F-4. My backseater sounded amazingly calm as he called off the altitudes as we spun through them. Our Phantom had "departed." (A departure meant that you had left controlled flight for uncontrolled flight. Simple, huh?)

I got the rotation stopped and tried to push the nose farther down. We were supposed to bail out at ten thousand feet if I had not recovered the jet. We had about two hundred and fifteen knots indicated airspeed. I pulled back gingerly on the stick. It flopped back to its rear stop like it wasn't even connected to the stabilator. I shoved it forward again. We were passing ten thousand feet. I could see nothing but blue water through the windscreen. I pulled back on the stick again. Nothing! Our pitch angle had not changed one degree. The sea was getting closer.

I had to do something. I immediately rammed the throttles into full afterburner. I could feel them as they ignited. The thirty four thousand pounds of thrust shoved the airspeed indicator off of its stagnant reading. At two hundred and eighty knots, I pulled back on the control stick again. The stabilator responded as designed, and the nose rose rapidly. We leveled off at about three thousand feet.

Even *I* thought my reaction was strange. I was pissed! Not because technically I had been bested in a dogfight, but because I felt that my Phantom had betrayed me, and I didn't know why. As soon as we landed, I went to the flying safety office to try to figure out what had just happened.

After my report, the safety representative I talked to said he would look into it. We had recently lost one F-4 in a "departure" accident during an air-to-ground bombing run as it was rolling in. The pilots got out. We had also lost the Phantom on the Florida Avon Park gunnery range right after we had received the planes. Those pilots did *not* get out. The Navy had crashed some F-4s in unexplained loss-of-control accidents. Other Air Force F-4 units had lost aircraft due to loss-of-control (departure) accidents.

It is my belief that my particular Phantom, in its combat configuration that day, was in an aft CG situation. (An aircraft needs to be in a condition of aerodynamic "balance" when it is in flight. If its center of gravity is too far forward, it is stable but nose heavy. If the CG is too far back, it is tail heavy and very unstable.) I had about four thousand pounds of fuel remaining, all in the fuselage tanks. I was carrying a TER (triple bomb ejector rack) under each wing on the inboard station. I had an empty three-hundred-and-seventy-gallon fuel tank under each wing. I suspect that one thing probably put the Phantom's CG aft of the normal limit: the centerline Gatling gun with six hundred rounds of 20-millimeter ammunition remaining.

I believe the combination of the half-loaded gun and distribution of the fuel remaining in the fuselage tanks almost "got me." The jet was placed in an angle of attack and airspeed where the stabilator was totally ineffective.

(I'm not sure whether my incident and report had anything to do with it or not, but the F-4Cs I flew in the Air Guard in later years had an interesting modification on the fuel panel. It was called a "5-6 lockout switch" that allowed the fuel-feeding sequence to transfer fuel out of the rear tanks forward into the feeder tank much sooner, which resulted in a more-forward CG.

One thing *is* for sure. Being a typical fighter pilot, I sure as hell wasn't going to believe that I had done anything wrong to cause the Phantom to "depart from controlled flight." Damn betcha!

Postscript. When I got to my squadron, there was a Red-Cross telegram awaiting me. I was the father of a beautiful baby girl, Kristin

Lynn. There's a saying that "Fighter pilots usually sire girls." The theory is that it's the Gs.

The thud, the brunt

North Vietnam, 1965–1966

It was designed for an ultra-high-speed low-level nuclear-bomb delivery vehicle. It was exceptionally heavy for a single-engine aircraft. It was fast, around Mach 2.2 at higher altitudes. It would outrun just about anything on the deck. It could carry six tons of conventional weapons. It was the workhorse of the fighter-bombers and flew about seventy-five percent of the missions into North Vietnam. It was the F-105 Thunderchief, more commonly known as the "Thud."

The F-105 pilots were respected by all other pilots, especially those of us flying in Southeast Asia. Their valor and performance in the face of overwhelming defenses and odds were fast becoming legend.

One of these pilots had been my working partner on a brief project in Saigon. While working during the day with him at Tan Son Nhut Air Base and having meals with him in the evening at the hotel, I noticed there seemed to be a "wall" around him that was almost visible. I know now he had placed it there to protect his emotional well-being. If I tried to get him to talk about the missions he had flown into North Vietnam, he would not say very much. It almost seemed like a superstition. I respected that and stopped asking questions.

He had to go back to Thailand and start flying to North Vietnam when we finished our temporary duty. The odds against him finishing his one hundred missions without being shot down were very high. He was a seasoned combat pilot and quietly resigned to his fate.

I still don't know what that fate was.

While flying MiGCap and escort missions into North Vietnam, I could hear distinct differences in the voices from the various cockpits. The tones and emotions of the pilots directly reflected their experience level and their attitude. I did not personally hear the following radio conversations among these F-105 strike aircraft, but they became rather well-known and repeated often. I am relating them as I heard the accounts.

The four-ship of F-105s was in combat spread formation and ingressing their target area to the west of Hanoi. As they got closer to their heavily defended target, the enemy defenses began to intensify.

"Lead, this is Four," came the urgent sounding call.

"Go ahead Four," replied the rather casual sounding flight leader.

"Roger Lead. They're shooting at us from our three o'clock," called the excited wingman.

"Ah, roger Four. You do know that they're allowed to do that?" came the laid-back question.

With nearly the same scenario and in the same general area of North Vietnam, the following was heard.

"Lead, this is Four."

"Go Four," called his leader.

"Sir, I got slung out wide on your last turn away from me, and I've lost the formation," came the obviously inexperienced and worried wingman's apologetic call.

"Do you know where *you* are?" came the now-concerned flight leader's question.

"Yes sir," came the very hesitant answer.

"Well, *where*?" the obviously exasperated lead F-105 pilot queried.

"I'm over downtown *Hanoi*," came the excited voice.

"**What the *Hell* are you doing there**?" the flight leader's question boomed over the air.

"*A thousand miles an hour!*" came the young fighter pilot's appropriate and probably accurate reply.

Again it was a four-ship formation of F-105s. They were nearing their initial point for their "pop-up maneuver" to visually acquire their target. It was time for a fuel and weapons check.

"(Scotch) flight, lead has switches hot and nine thousand pounds" of fuel remaining.

"Two has switches hot and eighty-eight hundred pounds."

"Three has switches hot, ninety-one hundred."

"Four has switches hot and, *I've got four thousand pounds!*" came the distressed Number-Four pilot's call. His voice had climbed several octaves, and it sounded like a fourteen-year-old girl's.

"Four, get the Hell out of here. Head toward the tanker, and climb to cruise altitude as soon as possible. And stay *out* of afterburner!" ordered the startled flight leader.

The seasoned lead pilot already suspected the cause for the inexperienced wingman's low fuel problem. The brand-new combat pilot in the Number Four F-105 had been using his afterburner to keep from getting behind the rest of the flight when they turned away from him during the extended low-level flight into the target area.

That tactic had increased the fuel consumption tremendously and caused the critical condition in which he found himself. Over North Vietnam was not a good place to be low on fuel.

Word was that he made it to the tanker and lived to fly another day, *out* of afterburner.

Another four-ship of "Thuds" had safely completed another mission to the north and were entering the traffic pattern at their base in Thailand. They were in a right-hand echelon formation as they made their steep-banked left turn onto the initial for their fighter "break."

Just as they rolled out of their turn to the straight three-mile run toward the end of the runway, came this call.

"Lead, this is Four."

"Go ahead Four," said the bored-sounding flight leader.

"Roger Lead, Three just *hit* me," called the perturbed-sounding Number Four F-105 pilot. (Evidently, the Number Three "Thud" had bounced a little and number four was up just a little too tight under him during the steep turn and roll out onto initial. They had actually banged together slightly.)

Back came the leader's authoritative command, "Three, stop hitting Four!"

"Roger Lead," came Three's call. They landed with no more "hitting."

The purpose for an old former F-4 pilot repeating these F-105 "Thud" stories is to demonstrate the strength of the human spirit under some of the most trying circumstances known to man. I know that you guys flew the toughest missions and took the brunt of the heaviest air defenses since World War II.

I know that most of you loved your airplane as I loved mine. I returned from two missions with one engine failed. You began with only one. You flew into "harm's way" on a single engine knowing that the odds were against you. You went regardless of them. You sustained your "cool" in any way that you could, and it was reflected in these anecdotes. I trust that they will serve to bring "home" to the reader your calm command of your reactions and your emotions under extreme circumstances.

I chose not to relate any "Thud" accounts with sad endings; you have too many of those to remember already. You're some of the most courageous guys in history. I salute you!

The longest day

Laos-North Vietnam border, 1966

It began as an escort mission for an EB-66 electronic-surveillance-and-radar-jamming aircraft. They were placed "on station" to help pinpoint and try to confuse the enemy radars. This would help the strike forces, normally F-105 Thunderchiefs, during ingress and egress of the target areas. The EB-66's "racetrack" flight pattern took it from about thirty or forty miles northwest of Hanoi to the area of Dien Bien Phu, near the western border of North Vietnam.

We were a flight of two Phantoms. We were each armed with four radar-guided Sparrow missiles and four heat-seeking Sidewinders. It was rare for a North Vietnamese MiG to make a try for an EB-66. The 66s were relatively slow and vulnerable, but the MiGs usually left them pretty much alone. It was early in the air war, and they were still wary of mixing it up with a couple of missile-carrying F-4s.

The flight was supposed to have been completed in less than four hours. We had been relieved of our escort duties and were headed back south for a hot meal. The day had been basically uneventful to that point, at least uneventful for us.

It had started with a shootdown of an F-105. The pilot was on the ground southwest of Vinh, North Vietnam, near the Laotian border. He was talking to his potential rescuers on his survival radio. The area of the shootdown had proven to be crawling with enemy soldiers and defenses. The rescue force aircraft had been driven off several times by ground fire. An airborne command post had called in a flight of aircraft to try to soften up the enemy defenses and effect the rescue. If I'm not mistaken, it was two B-57 Canberra bombers.

On one of their first passes, one of the Canberras was flamed and knocked out of the sky. I believe that one of the two crewmembers ejected, but the other was lost. Now there were two aircraft shot down, two crewmembers trying to evade capture, and one dead.

The airborne command post called us and requested our position and fuel state. We had just topped off with fuel from a KC-135 for our flight back to Cam Ranh Bay, so we were "fat." We informed them of that fact and that we were "MiGCap" configured, eight missiles and a centerline fuel tank. I don't know if they were concerned with enemy airborne interference with the rapidly deteriorating scenario or why, but they gave us some coordinates north of the shootdowns and told us to set up a standard MiGCap pattern between the rescue area and the MiG airfields to the north. We were all "working" the same radio frequency, so we were able to keep abreast of the developing situation. It was approaching late afternoon.

Two Navy F-4s were called in next for the continuing effort to silence some of the guns that were stopping the rescue attempts. On the very first pass, one of the battleship-gray Navy fighters suffered a direct hit. A raging fire engulfed the Phantom; the RIO and the pilot ejected. Situation: three aircraft shot down, four aircrew hiding or running for their lives, and one flier who had already lost his. Things were "going to Hell in a handbasket."

I must admit that as brave as I fancied myself to be, I was slightly "shook" at the rapidly deteriorating situation and not just a little bit glad that I was sitting here at altitude with air-to-air weapons. The way it was down below, I would rather have tangled with a half dozen MiGs all by myself than go down into that buzz saw.

I was southbound in my orbit and was scanning the ground below trying to see any of the activity. A CH-3 "Jolly Green Giant" rescue helicopter was trying to pick up one of the F-4 crewmembers. They had managed to get their Navy Phantom a little farther away from the concentrations of guns and the chopper pilot thought that he could get them out. As I watched, two of the A1E Skyraider helicopter escorts, called "Sandies," were trying to silence a gun that was close enough to reach the low flying helicopter. They appeared to be strafing with their 20-millimeter cannons. One of them pulled off of his weapons pass trailing smoke. As I watched, he climbed away, and then I lost sight of him. I heard one of the other pilots say that the pilot had bailed out, but that his chute did not appear to open. It had not. Situation: four aircraft shot down, two dead crewmembers, and still four evading capture on the ground.

There is bravery and then there is *bravery*! The next sequence of events is a real hero story. The Jolly Green Giant chopper finally positioned himself over one of the downed Navy F-4 crew. The guy had been injured severely in his ejection or parachute landing and could not get into the rescue sling unassisted. An enlisted paramedic saw his terrible plight and without hesitation went down the rescue cable to the ground and helped the wounded man into the sling. They hoisted him away to safety. The paramedic was in radio contact with the chopper pilots. Suddenly shots began ringing out around him. He told the chopper to "Get the hell out and leave me!" Then he took off running.

Situation: four aircraft shot down, two dead, one crewmember rescued, and *still* four on the ground evading the enemy.

The Jolly Green kept trying. Finally, they managed to pick up the other F-4 crewmember. It was getting dark. It was time to go. They withdrew with their two rescued guys and headed back to their home base. It was difficult. They had to leave three of their American brethren on the dark ground trying to avoid the enemy soldiers all through the night. That included one of their own, the paramedic who risked his life to descend into that ultradangerous situation for someone he didn't even know.

We were released to return to Cam Ranh Bay. I felt like some sort of traitor leaving those guys on the ground behind me, but there wasn't a thing that I could do for them except pray. I felt guilty going back to my own room and sleeping on a mattress. They wouldn't be sleeping at all. They would be hiding or running for their lives all night. Or they would be killed or captured.

A monsoon storm was right over Cam Ranh Bay when we arrived, and it didn't appear to be going to move elsewhere anytime soon. We didn't have enough fuel to wait around, so we headed to Phan Rang, another ten-thousand-foot-long aluminum runway about twenty miles south of us.

We finally found some rations to eat. It had been a long time since breakfast. We ended up sleeping on the floor of someone's tent. It was a lot better than those we left behind. I felt *guilty.* I still do.

We finally got some word on the rescue attempt the next morning. I heard that they got the F-105 pilot out. He had been the first one shot down. I still don't know about the rest.

I found out a few days later that the Skyraider pilot whose parachute hadn't opened was an old friend of mine from Laredo AFB. We had been instructors together in the same flight. Another dead friend.

Wayne's Waterloo (almost)

Coastal Lowlands, RVN, 1966

Wayne not only was one of the best fighter pilots I ever flew with, he probably looked just like most people's idea of how a fighter pilot should look. He was about six feet of tanned one hundred and eighty well-conditioned pounds and could have played a fighter pilot in the movies (à la Tom Cruise in *Top Gun*). He was good. His flying techniques were smooth and precise, and he could get the very maximum out of his F-4.

Some pilots in fighters are what I call "bankers and yankers." Their control inputs are abrupt and jerky. They think that they're max-performing their fighter when, in actuality, aerodynamic performance is being lost by the disrupted airflow caused by such a rough technique. Certainly there are emergency situations requiring instantaneous control inputs (missing the ground, for instance), but they do not occur often. The pilots I am referring to fly this way as a matter of routine and think that makes them a "fighter pilot." Nope! It makes them "ham-handed" and very hard to fly formation with. Most pilots fall somewhere in between these two extremes. Wayne didn't. He was at the top of the ability and skill spectrum.

I was on five-minute alert at the south end of the runway as the number-two aircraft in a flight of two. Wayne was number one. The scramble-phone rang, and we were airborne in about three and one-half minutes. Our target location turned out to be in the coastal lowlands approximately one hundred miles north of Cam Ranh Bay. The South China Sea was in sight on our right as we descended toward our waiting forward air controller.

The FAC caused us to be scrambled. He had obtained fresh intelligence and definite indications that a Viet Cong command post had been established in a large white stucco-type building with a reddish tile roof. The building was located on the northeast edge of a small town. He also had information that the basement or bottom floor of the building was being used for ammunition storage, but there was no way to verify it. His sources had been monitoring activities in and out of the building, and there were reported to be a large number of high-ranking Viet Cong inside at this time. He wanted to catch them there before they left, which was the reason for our scramble from alert.

We decided to use our Gatling guns for the attack because of the proximity of other structures. The guns were extremely accurate, and we could contain our attack area.

The building was so prominent that the forward air controller did not have to mark the target at all. He just described it, asked a few questions to ascertain our positive identification of it, and got out of the way.

Wayne had initiated a fuel and weapons check as we had made our wide orbit of the area. We were ready. Wayne positioned his Phantom for a left-hand strafing pattern and rolled nearly inverted as he lined up his gunsight. Meanwhile, I was about ten seconds from rolling in for my first past. As I watched, the familiar smoke trail began streaming from beneath the belly of Wayne's F-4. The smoke was the result of one hundred rounds per second of twenty-millimeter cannon being fired. I observed the first high explosive incendiary rounds strike the bottom of the near wall of the white building. The bottom part of the wall disintegrated as the bullets walked up the side toward the second floor.

Wayne stopped firing, and I saw the vapor trails begin tearing away from the top of his F-4 as the low-pressure areas forming on top of his wings instantly condensed the humid coastal air. The nose of his jet continued rising, and then it happened. The building disintegrated in an enormous fireball, except for the big red tile roof. It rose, almost intact, directly up into the Phantom's flight path. What immediately followed was one of the most amazing things I've ever seen.

Wayne's Phantom couldn't clear the roof. Instead, it roared into the fireball just below the red tiles. My heart leaped into my throat as I witnessed what I thought were the deaths of two close friends.

Suddenly the charging F-4 emerged miraculously from the far side of the fireball. Its nose was up and it was climbing like a "*bat out of hell*," which is a pretty accurate metaphor for what I was seeing. To make it even more spectacular, fire from the exploding building was chasing up the twin-wing vortices behind the Phantom. They resembled miniature red tornadoes. They couldn't catch the charging F-4. I don't think the devil himself could have caught it.

We joined up on Wayne's wing and looked his jet over for damage. We saw hardly anything unusual! How he could have flown through all that fiery debris with only the minor damage found later on the ground is amazing. It might not have been entirely attributable to Wayne's excellence as a fighter pilot this time. Ever seen a miracle?

By the way, Wayne and his backseater were credited with confirming that the building had indeed been an ammunition storage area!

Me? I never fired a shot.

A most unlikely savior

Cam Ranh Bay Air Base, RVN, late summer 1966

About three quarters of a year gone, a few months to go: I was playing table shuffleboard in the Officers' Club bar. There was lots of noise and laughter as always. The fighter pilots were unwinding after their day at the office. Some of the bachelors were flirting with the Red Cross girls and nurses. The ratio of men to women was five hundred to one at Cam Ranh, so the competition was always pretty fierce for the females' attention. The sky outside was dark and overcast, and all the day's missions had apparently been accomplished; at least there had been no flying activities for an hour or so.

The clouds lit up brightly to the north. Before we could even think about what the bright flashes had been, the sound of an explosion reached us. Shortly thereafter, it was followed by an even louder explosion. At first, most of us thought that we were under some sort of attack, but there were no sounds of gunfire and no trip flares going off. Some of those present returned to their previous activities, but I and several others decided to try to find out what had happened. As I walked toward the flightline, I heard a helicopter take off. It proceeded to the north, which had been the direction of the flashes and explosions. When I arrived at our base operations building, several others were already there. When I asked what was going on, they said that the word was an aircraft had exploded over the water to the north of the base. They also said a search for survivors had begun.

The helicopter was circling low over the water a few miles north. It stopped and hovered in place from time to time. Suddenly, it made a beeline back to the base and landed. We walked out toward its position, and they were transferring someone from the chopper into an ambulance, which roared off as the helicopter took off toward the north again.

Slowly, the facts unfolded. There had been two F-4s, both from my squadron. It had been a long day for the four pilots. They had been diverted from their first target because of weather. When they had been unable to get into their secondary target area, their fuel state had forced them to land at Da Nang Air Base for refueling. After some delay, they finally received their fuel and took off heading for Cam Ranh Bay.

The aircraft commander in the Number Two Phantom had just transferred control of the aircraft to the backseater so that he could check a switch position in his cockpit. They were in normal formation on lead's left wing. Evidently, just as the aircraft commander in the second Phantom lowered his head to check the switch, the lead F-4 turned left to begin an instrument descent toward the east. The young backseat pilot, who I believe was only on his second mission in Vietnam, didn't react in time. They collided in midair (the first explosion).

As the frontseat pilot in the second aircraft felt the collision, his cockpit filled with burning jet fuel. He was covered with fire. He grabbed the ejection handle and pulled it, but his aircraft had already disintegrated around him. As the stricken F-4s fell away, he was still burning. Both aircraft were still carrying the five-hundred-pound bombs that they had been unable to drop earlier. One of them evidently detonated (the second explosion).

Not one piece of shrapnel from the bomb struck him, but the concussion did. The fire was snuffed out. He couldn't recall anything after that, but he was the only one that the helicopter found.

The other three pilots were gone forever.

Postscript. The lone survivor not only survived, he went on to recover and became a pilot for a major airline, thanks, perhaps, to a bomb.

The elephants versus the ants

Cam Ranh Bay Air Base, RVN, 1966

"He's stomping 'piss' ants while the elephants are trampling all over us." I had heard that expression ever since I had entered the U.S. Air Force. It referred to anyone who was consumed with small relatively unimportant details when larger more important things were being overlooked. It has always amazed me that so many otherwise intelligent people seem to fall prey to this fact of life.

The word had started to filter down a little at a time. Things were going to change. We had a new general in Saigon at Seventh Air Force. He was making his presence known with some sweeping policy changes. Lieutenant General Moore had been the Second Air Division Commander, which became Seventh Air Force, since I had gotten "in country." He was another fighter pilot's fighter pilot. He was much admired and liked by all of the combat pilots and aircrews who knew anything about him. But good things always seem to come to an end. General Moore was leaving. His tour was finished.

In combat zones, things cannot be exactly as they are in peacetime, nor should they necessarily be. Any good officer or NCO (noncommissioned officer) knows the absolute necessity of good morale; one who doesn't has a real problem with reality.

It is really amazing how little it takes sometimes to attain and maintain good morale. How about letting the troops wear an Australian "Go-to-hell" hat if they want? What about letting the aircrews wear their locally produced western-style gunbelts instead of the standard issue holsters under their arms? Maybe the guys could let their mustaches grow just a little longer. How about rolling up flight suit sleeves and unzipping the suits halfway down the chest because it was so damned hot? Do those sound like a big deal in a war zone?

The first victims were the stop signs on the base. If my memory serves me, they were white letters on red backgrounds. Man, you just can't have that kind of gross insubordination! Those signs *have* to be changed to black on yellow backgrounds! That was important stuff! We changed the signs.

"You *will not* wear any unauthorized headgear." Away went the "Go-to-hell" hats. (Don't ask *me* why they were called that. Ask an Aussie.)

"You *will not* wear western-style gunbelts. You *will* carry the .38-caliber Combat Masterpiece in the holsters provided for that purpose." The gunbelts ended up hanging on nails in our rooms. Good military men obey orders.

"You *will* trim your hair and mustaches in accordance with Air Force Regulation 35-10!" We were *really* having fun by now. Morale was skyrocketing!

I was wondering how in the world any of us ever got to the flightline with stop signs of the wrong colors. How in hell had an F-4 ever been able to get airborne on a combat mission with its pilot's mustache out of limits. I guess we had just been lucky up to now.

"There will be a meeting of all available aircrews in the wing briefing room at 0800 hours tomorrow morning. A representative from Seventh Air Force will be here to explain new operational procedures that will be taking effect immediately," so read the posted sign.

I was airborne the next day before 0800, so I did not get to hear the higher headquarters briefing. But I sure heard about it!

The major was officially representing the new "ant stomping" general. The only part I heard about the briefing affected me exactly the same way that all the other pilots had reacted. According to some of those who had been present, it went something like this: "From now on all pilots leading combat missions out of bases in South Vietnam will file flight plans with Saigon via land line. It will include callsign, number of aircraft, takeoff time, route, altitude to and from the target with the coordinates of the target, and the time on and off target," the briefing major said.

There was a tremendous shocked silence according to my source. There then began much headshaking and a rising murmur as the fighter pilots conversed quickly among themselves. I don't know who instigated it, but supposedly they all stood up almost at the same instant and began heading toward the exits. The surprised briefing officer raised his voice in protest and said that he wasn't finished yet. Someone yelled back toward the stage, "You're finished!"

"But you can't just leave. What will I tell the general?" he protested. I wasn't there so I can't say for sure, but apparently, almost in unison, the answer came back from the fighter pilots, "**F— You**!" They walked out.

We had already been suspicious from time to time that the Viet Cong were somehow intercepting our target information. Based on the apparent value of some of our targets, I sometimes wondered if the Cong weren't *picking* them! Yeah, that was really what we needed to do! File a flight plan to help them be ready for us when we arrived!

We never heard any more about filing combat flight plans! In fairness to the general, this *all* could have been the brainchild of some staff officer trying to impress him. The general's decisions could only be as good as his information! On the other hand, remember the "No F-4 accidents" general? Same guy!

I did learn something very valuable from all this.

"Sometimes it's better to have certain individuals stomping piss ants and leave worrying about the elephants to the rest of us!"

"White Anchor, break left!"

Offshore, North Vietnam, 1966

You'll recall that the U.S. Air Force had finally gotten enough Air-Refueling KC-135s in the Southeast Asian area for us to start flying into North Vietnam. Because Cam Ranh Bay was about 350 miles south of the DMZ between North and South Vietnam, air-refueling tankers were essential.

The other F-4 in my flight of two and myself had been gone from our home base for several hours. Two aircraft out of the four that had been scheduled to relieve us had not shown up. That meant that we had to stay twice as long as originally planned. The mission we were assigned required two F-4s to remain with the KC-135 orbiting at the White Anchor refueling track while two more were flying fighter cover for an EC-121 radar plane. The EC-121 was to fly up and down offshore from North Vietnam and maintain a constant radar, radio, and other electronic surveillance of the enemy's aircraft and surface-to-air missiles (SAMs). This was a vital mission in 1966 because the electronic warning equipment installed in fighters later was not yet available. An alerting call from the EC-121 was all the warning, other than visual sightings, that a fighter pilot had over North Vietnam.

The EC-121 was effective in its mission, but very slow and vulnerable; therefore, our mission was to protect it from enemy fighters. That required two Phantoms to be in close proximity to the EC-121 at all times while it was on its racetrack flight pattern. The other two would top off their tanks and fly north to relieve the two F-4s protecting the radar plane as they approached a preplanned minimum fuel remaining number. Then they would head for the tanker until they were needed again.

I believe that we had been airborne about three and a half hours when our relief four-ship showed up as a two-ship. Well, rank has its privilege, so Willy and his backseater, and I and mine, were still "in the barrel" so to speak. That meant we had about another two hours or so before we could go "home."

Finally the time came, and we were allowed to start toward South Vietnam. Our tanker had been relieved also, so we just flew with him as he headed back south. We needed one more fuel top-off, and Willy and his backseater eased into refueling position and began to take their programmed fuel load. I sat off to the right side and just tried to relax and find a more comfortable position. Anyone who has flown an F-4 knows how futile that is. Willy got his gas and dropped back and to his left. He joined up loosely on the tanker's left wing as I began to move left toward the precontact position. As I stabilized, I glanced down on my right. I could see the water of the Gulf of Tonkin below us.

I looked to my left and saw that the backseater was flying Willie's Phantom. It was obvious because Willie was attacking what was left of his flight lunch in the front cockpit. The boom operator gave me the light signal indicating that the tanker was ready, and I slid forward toward the air-refueling boom. It took quite a while to get my fuel load. We still had a long way to go and wanted our tanks full. As I slid back and down following disconnect, I glanced down to close my refueling door switch. I then looked over the left canopy rail, and I saw land! We were not supposed to be over land yet! I banked hard right and looked again. We were directly over Vinh, North Vietnam.

"*WHITE ANCHOR, BREAK LEFT!*" I yelled over the radio. The pilot of the big KC-135 began rolling into a left turn as rapidly as he could. I frantically looked over at Willie's F-4. He was in a hard left bank to stay away from the tanker.

"Roll out on a heading of zero nine zero degrees, and get the hell out of here!" I yelled again. The KC-135 was adding power to accelerate, and I let him go. I was busy looking down at the ground for surface-to-air missiles or big antiaircraft gunfire. I also figured that right then, the farther away from him I was the better.

"What's the matter, Whiskey?" the tanker aircraft commander asked.

"We're directly over Vinh, North Vietnam, right now and they've got missiles and big guns down there," I replied.

I continued to zigzag, and Willy was doing the same. The tanker was by now hauling "you know what" out of there toward the beach.

We made it.

I cannot understand to this day why the North Vietnamese didn't shoot. Maybe they were too busy trying to believe their eyes until it was too late. We were just lucky. One shot could have "hosed" all three of us while we were refueling.

"Whiskey, thanks for the 'heads up.' This is our first mission, and I guess we just got lost," said the KC-135 pilot.

I wanted to be kind, but I just couldn't: "You keep that shit up and it'll be your last!"

Willy waggled his wings, and I joined up on him for the trip to Cam Ranh Bay. There was chocolate milk all over the inside of his canopy. But I guess that's better than a missile up your butt!

Postscript. On the way home, while about halfway between Da Nang and Cam Ranh Bay, we decided that we wanted to go fast. After all, we had already had a long day. Minimum afterburner seemed like a good idea, and we didn't even have to worry about filing a sonic-boom report like we would have in the states.

I don't remember how fast we were going, but it was supersonic. I guess my Phantom had also gotten tired during the long day and decided to turn on its right-hand engine-fire warning light. (It always seemed to be on my right.)

I pulled it out of afterburner, but I didn't want to slow down that much. I simultaneously pushed up the left-engine throttle into a higher afterburner stage to partially make up for the loss of the right one. The engine fire light had not gone out yet, and Willie dropped back to look me over. I lost sight of him as he did. I don't remember if we were still supersonic, but I think so. Finally, Willy called me in his calm Arkansas drawl.

"Jerry, there's no sign of fire, but now I've got just one engine running. I don't know why you did that to me when I'm just trying to help!" he laughed.

"Do what?" I asked.

"Well, when I pulled up behind you to look up your tailpipe, *you* blew out my left engine," he stated rather matter-of-factly.

"I'm sorry. Why don't you start it up, and let's go home?" I suggested.

"Already did," Willy replied as he joined on my wing.

We went home. I seem to remember I had to promise to buy his dinner and some more chocolate milk to shut him up. He blamed me for spilling his all over the cockpit a little earlier that day.

Missiles (useless as teats on a boar hog? *Sometimes*!)

Cam Ranh Bay, Air Base, RVN, 1966

We were getting to fly more and more missions into North Vietnam and over her coastal waters. We were hunting MiGs while protecting electronic-countermeasures aircraft and providing MiG cover for our fighter bombers. I was glad to get the opportunity to be in "MiG country," but I felt somewhat "hamstrung" as did most other F-4 drivers in 1966.

The United States Navy had shown great wisdom when it purchased the F-4 Phantom II from McDonnell Aircraft in the late fifties and early sixties. The F-4 had been an amazing leap in aircraft performance over previous fighter designs. It was to join the F-8 Crusader as the Navy's first-line fighters. Although the F-8 was a beautiful and fast Navy and Marine fighter plane, it was essentially a day "gunfighter" designed for close-in air combat. It had been modified to carry sidewinder heat-seeking missiles, but basically required a visual acquisition of the enemy fighter plane and utilized a dogfighting-type combat scenario. This was certainly a most agreeable way of fighting to every fighter pilot *I* have ever met. It was essentially the way air combat had been fought since World War I.

However, the Navy had seen a need for a fleet defense interceptor, and the F-4 was it. The idea was for the fleet radars to see the enemy planes inbound and launch the F-4s against the threat. The F-4's radar would then acquire the targets, and its radar-guided Sparrow missiles would be launched against the enemy aircraft. The fleet's Phantoms could theoretically shoot the enemy fighters or bombers down without ever seeing them; therefore, the F-4s would not need a gun. Consequently, they didn't have them.

According to the latest conventional wisdom of the time, air combat would be changed forever. It would be strictly radars and missiles. The days of the gunfighter were over. Right? *Wrong!* Every fighter pilot knew better, but no one listened. The "brass" making the decisions thought that the fighter pilots were just grousing because they were trading in a "fun" mission for a dull "by the numbers, center the firing dot on the radar and shoot when the backseater told you to" scenario. Admittedly, that was part of the reason that fighter pilots wanted a gun, but the legitimate reason was the very *real* problem of aircraft identification.

There was no way in that era to be sure without seeing the unknown aircraft. We knew that no commanders were going to authorize the firing of missiles at an unidentified target until it was definitely verified as hostile, and that meant a visual sighting by the aircrew. When our "missile-only" aircraft had gotten close enough to accomplish the identification, they were usually too close to fire a missile. The missiles had very narrow parameters for successful firing and needed a minimum distance to arm. Not only that, those parameters were constantly changing because of the fluid dynamics in a three-dimensional high-speed conflict.

Seems like a simple-enough concept to understand, doesn't it? All of the fighter pilots wearing bars and oak leaves understood it. Why didn't the admirals and generals wearing stars appreciate the obvious? Probably they did! More than likely it was just simply civilian and military politics raising their ugly heads again. Anyway, the Navy didn't get the damn gun and neither did the Air Force when we first bought the Phantom. Later Air Force versions did come equipped with an internal Gatling gun, *finally*.

The "missile problem" was somewhat like trying to fight someone who had a knife in a five-by-five closet when your only weapon is a ten-foot spear.

In my humble opinion, the only fighter with the proper capabilities to fight the air war in North Vietnam—and with its pilots still properly trained to have a chance of winning on a regular basis—was the gorgeous little U.S. Navy F-8 Crusader mentioned previously. It could carry sidewinder heat-seeking (actually infrared energy) missiles but also sported an internal gun with a computing gunsight for the close-in "furballs" (air-to-air dogfights).

In flat-out performance, the F-4 could best anything the enemy fighters could do except turn (at slower speeds below 450 knots) or fire a gun (which, of course, it didn't have). *If* F-4 "drivers" had been properly trained in air-to-air combat to utilize their Phantom's advantages in certain situations, the air-to-air story of Vietnam would have been much different, even without a gun.

(Proper, effective, and realistic air combat training was *gravely* lacking in the early sixties. You *know* the Air Force and Navy both felt they couldn't take a chance on losing an airplane just out there doing that foolish dogfighting practice! The folks in power were convinced that close-in air battles were outdated nonsense anyway. *Now* we had **missiles**!) "Yeah—Sure!"

As the fighter pilots suspected it would be, the air-to-air kill ratio over North Vietnamese MiGs was miserable. It was on the order of one U.S. fighter lost for every two enemy planes downed. In Korea, it had been something like one American plane lost for every thirteen enemy MiGs downed.

Imagine if you will, a heavyweight professional boxer's manager and trainer not allowing him to ever practice sparring because they're afraid that he'll get hurt. I can just hear them: "Just go read a book on old-fashioned boxing, George. *We* think they're going to change the 'rules of boxing' to 'standing a thousand yards apart and throwing rocks.' Of course, you'll have to take your boxing gloves off to 'chunk' the rocks. Oh, and by the way, you can't throw any rocks at them until you can definitely identify them as your opponent for sure. Oh, and it might be cloudy or dark when they want to fight. And if they happen to get in too close to you to throw, you can't hit them anyway because you don't have any boxing gloves. Sorry! Oh, and you'd better win or everyone watching will think that you're not any good!"

We were the heavyweight boxers, and the generals were the trainers. The managers were politicians and government "drones," such as our Secretary of Defense at the time.

Enough "sour apples." There's a happy ending. I'm very pleased to announce that Vietnam "bred" some combat-experienced "fighter" generals who not only had been there, but knew how to *listen* to the troops.

These days, our fighters not only "turn on a dime," they have missiles *and guns.* Oh, *and* the pilots are allowed to train in air-to-air combat.

It might be old-fashioned, but it's still necessary, and fun! I miss it every day.

Famous, almost (infamous, so close)

Gulf of Tonkin near Haiphong, 1966

My desire had been much greater, so far, than my opportunities. I had declined two R and R leaves that I had been offered to go "hunting." I wanted a MiG. They just weren't coming up to fight. I flew on every mission I could that required air-to-air missiles and was either a MiGCAP or fighter escort. To date, I had only spotted two MiGs visually and a probable on radar but had been driven off by SAM warnings every time before I could get in missile range. I wanted a MiG, badly.

The radar operators on the EC-121 that we were escorting were clearly nervous. They obviously had heard the same information that we had received in the morning's intelligence briefing. The pitch of their voices and the urgency of some of their radio transmissions were dead giveaways.

I didn't blame them. There they droned, a few miles off the coast of North Vietnam at an airspeed that my Phantom fighter only slowed to just before touching down on the runway. The huge radar antennas on the triple-tailed Convair Connie slowed it even further. They were there performing an essential job for the Air Force and Navy. In 1966, the technology had not advanced to the level of a few years later. We had no electronics in our fighter cockpits, except our radios, to warn us of enemy fighters or surface-to-air missiles. The EC-121 had equipment to detect radar emissions and aircraft. A call from the EC-121 or another aircraft of similar capabilities was all the warning we had, except for our eyes. With restrictions to visibility such as clouds, fog, or smoke, these warning radio calls were vital. A SAM launching toward you on a clear day might be seen if you happened to be looking in the right direction, but we didn't particularly like depending on just our good luck to keep us from getting shot down.

The intelligence briefing that morning had included a leak from a usually reliable source that the North Vietnamese Air Force had decided to try to sneak a MiG fighter out over the gulf to shoot down the EC-121. It had become a thorn in their side. They believed, and rightly so, that it was directly responsible for lowering their success rate in shoot-downs of United States combat aircraft. *My* mission was to prevent the successful completion of *their* mission.

As I stated, the EC-121 was relatively slow and unmaneuverable. Its only hope of protection was its fighter escort. I was one of four escort F-4s that day. I was itching for some action as much as the EC-121 guys were hoping against it.

About ten minutes after an air-to-air refueling, I arrived on station and relieved two other Phantoms who then headed back to the KC-135 tanker. They would refuel and wait another turn at escort.

Five minutes later, "Whiskey 11, two fast movers, very low, vector two eight zero!" came the excited call. I immediately banked my Phantom to a heading of two hundred and eighty degrees and began to accelerate and descend. My wingman, callsign Whiskey 12, followed. As we rolled out on our assigned heading, I saw that it would take us almost directly toward Haiphong harbor, which was a hotbed of enemy defenses. My backseater was focused on his radarscope, which was scanning the sky ahead and below us.

The radar operator on the EC-121 reported that the two suspected enemy fighters, or possible "bogies" as they were called, had begun an orbit about five miles offshore and were still very low. They also reported that the unknowns were not transmitting the electronic code of the day to identify them as friendlies. I told my wingman to assume radar-trail position. That was several miles behind me. It was designed to put him in a position to fire radar-guided missiles at the aircraft if I could visually identify them as enemy then break away out of the line of fire. As I accelerated my F-4 and descended toward the water, I was in a quandary. I had already relinquished those two R and Rs to hunt MiGs. It certainly wasn't my inclination at this point to break away, if these were enemy fighters, and let my wingman have the kills. It didn't take me long to decide. That wasn't going to happen.

My backseater made radar contact with the unknowns and relayed that fact to the EC-121 and our wingman. By now, I had leveled off at five hundred feet above the gulf and adjusted the throttles to maintain six hundred knots. I glanced down quickly at my radarscope and saw the two radar contacts dead ahead of us at about five miles. I looked up and continued to try to acquire the two unknowns visually. I was having a lot of difficulty due to the appalling visibility from the smoke and haze.

"SAMs yellow!" came the warning call from the EC-121 to us. That meant that they had detected surface-to-air missile radars being activated in our area. To me, that seemed a strong indicator that the aircraft that we were intercepting were probably enemy; otherwise, I reasoned that the surface-to-air missiles would have been activated before we arrived because of *their* presence. My adrenalin was flowing freely now. My heart was pounding as I peered through the windscreen.

Suddenly, there they were! I saw a glint of sun off one of the aircraft. At our eleven-thirty position and slightly above were two fighters in about a thirty-degree left bank, heading away from us. I told my backseater over our interphone that I saw them. I was going about three hundred knots faster than they, so I extended the speedbrakes, pulled the throttles to idle power, and performed a very high-G barrel roll to the right to kill off some airspeed. As we rolled out behind the fighters, about a half-mile behind and below, I flipped my missile switch to heat and placed the center of my gunsight on the trailing fighter's tailpipe. Immediately, the selected Sidewinder heat-seeking missile began its insistent growl in our headsets. That meant that the Mach-Two-plus missile was ready to fire. As I strained my eyes trying to verify the identity of the two fighters in the poor visibility, they began to appear more and more like MiG 21s, North Vietnam's hottest jet fighter.

"I see them, they look like MiG 21s," my backseater excitedly called over our interphone.

"That makes two of us," I thought, and I reached down to my armament panel. As I flipped the arm switch to "on," I knew I was grinning. All I had to do now was squeeze the trigger, and I would have my first MiG kill. I could almost see the headlines in my hometown paper, "Captain Cook downs two MiGs over North Vietnam." I had waited and trained for this moment for years.

The visibility in the smoke and haze was terrible, especially toward the sun. I was excited, but I still wasn't one hundred percent convinced that these were bad guys; however, I was definitely leaning in that direction. I certainly wanted them to be!

"Whiskey 11, this is 12 . . . my radar's intermittent and I've lost contact," came my wingman's call.

"12, break hard and head zero nine zero," I instructed him. He "Rogered" and complied. The Sidewinder was howling in my headset as I closed the distance between myself and the possible enemy fighters. My index finger was straight out beside the trigger, but my remaining doubt wouldn't let me squeeze it. Suddenly, the fighters in front of me broke hard left and appeared to maneuver after my wingman, definitely a hostile move.

My doubt was rapidly vanishing. Just then, the Number Two unknown banked sharply to cross to the inside of his leader's turn, and I saw it! His wing-attach-point was at the top of the fuselage. MiG 21 wings were attached lower on the fuselage and were a different shape. I was simultaneously disappointed, relieved, and *very* agitated: disappointed that they weren't enemy, relieved that I hadn't fired, and angry at the two Navy F-8 Crusader pilots for not squawking the proper codes and for screwing around the way they were.

I told my backseater to change our radio transmitter to Guard channel, which all United States military aircraft monitored, and safed my missile switches. The F-8s had selected afterburner and were trailing my wingman, who had no idea what was going on. I knew that the F-8s wouldn't mistake Whiskey 12's F-4 for an enemy because no enemy aircraft laid down a black smoke trail like an F-4 Phantom. In fact, you usually saw an F-4's smoke long before you saw the aircraft. I joined closely on the second F-8's left wing. Just as I did, the anticipated "smart-ass" radio call on Guard channel came.

"Air Force F-4, fifteen miles east of Haiphong heading zero nine zero, better start looking around ace."

I punched the microphone button and said "Two Navy F-8s, trailing the F-4, better check your left wing."

The pilot of the nearest F-8, whose wing I was on, slowly turned his head and looked at me. I could see his head move as he obviously said a four-letter word to himself. I banked left to show him my eight missiles and rolled right back to his wing.

I punched the mic button again and said, "Now, why don't you get your butts back to your ship before you do something else stupid!" Without another word, the two F-8s turned south and began climbing. The last I saw of them, they were headed toward Yankee Station where their aircraft carrier waited.

I can't describe the mixed feelings that poured over me. I remember that the strongest one was the disappointment of them not being MiGs. I cannot explain why neither aircraft was transmitting the identification code of the day. I cannot explain why the SAMs around Haiphong hadn't activated at *their* presence. I don't know what they were doing there or why they made the aggressive move to chase my wingman. I do know how close at least one of them came to getting shot down that day! Neither one of us could have *lived* with that.

By the way, I never got my MiG, *or* an R and R.

Politicians and war (a deadly combination)

Cam Ranh Bay Air Base, RVN, 1966

I did not have a very high opinion of most politicians when I went to Vietnam. My year there did not improve that opinion. Picture yourself flying combat missions three out of every four days. Picture doing it on many occasions with partial loads of two-hundred-fifty-pound bombs and a half load of target-practice-ball ammo for your Gatling gun because you have run out of bigger bombs and real ammunition. (Ever hear the term, "hunting bears with a switch?") Now, send *four* aircraft with their eight pilots instead of the *two* aircraft that could have carried the shortened ordnance loads. Do that so that your branch of the service will not be shorted still further in the next appropriations by the Washington bean counters working for the Defense Secretary. Now picture the Secretary of Defense announcing to the American public that there is *no* shortage of bombs or ammunition in Vietnam.

Last but not least, listen to your President in a speech with his folksy manner saying, "My friends, we're not in Vietnam to win." Try to swallow a Commander-in-Chief saying that while you're getting *your* ass shot at and see how it tastes.

I wasn't in any other wars, but I think I am on safe ground when I say that this was probably the worst managed "war" in the history of the United States. There's one thing that I will never understand. Why do politicians place their country in a military conflict, and then second-guess and interfere with the professionals who spend their careers learning how to manage and win that conflict? How ludicrous. It's not the United States military that starts wars. I feel that many of our good citizens have a grossly distorted view of the military. They seem to think that military people love or want war. The farther left that their politics lean, the more inclined they appear to believe this fallacy. Only someone who has nothing to lose, or has something to gain, would want war.

Think about it. What has a military person got to gain? In my case at the time, it was an extra sixty-five dollars a month in combat pay and a five-hundred-dollar deduction on my income tax. No wonder I elected to go to war! *Right!* It was the money! What did I have to lose? Ask dozens of my friends and fellow pilots whose names are on "The Wall." Query their families. Try inquiring of the tens of thousands of other names on that black edifice. You'll find it very difficult to communicate.

Most of us had strong feelings about our President and Secretary of Defense. We didn't just suddenly acquire these feelings. That pair earned our feelings on a near-daily basis.

When we would go up against worthless targets that would be repaired by the next day and were hammered by heavy antiaircraft guns and missiles protecting a real target, such as a MiG airfield, we felt strongly about them.

When we were not allowed to pursue MiGs into the buffer zones around Hanoi and Haiphong, we felt strongly about them.

When we were not allowed to return fire that originated from within these buffer zones, we felt strongly about them.

When we would finally begin to learn where the enemy defenses were located so that we could knock them out, only to then have Washington suddenly call another bombing halt, we felt strongly about them.

When we would have to go back in after the bombing halts and find that all the defenses had been moved and replenished (from inside the off-limits Hanoi area and Haiphong harbor), we felt strongly about them.

When we were not allowed to attack the MiG airfields but had to wait to be attacked by those same MiGs in the air, we felt strongly about them.

Every time one of us was shot down while striking some valueless inconsequential target that those two self-appointed "military experts" had selected for us over their luncheon coffee, we felt strongly about them.

One of my fellow fighter pilots evidently felt stronger about them than others of us. It was early evening at the Officers' Club bar. This particular pilot was having a few quiet drinks by himself. Things were fairly boisterous as usual when he stood up and turned around. He raised his hands and shouted, "Quiet, I have an announcement to make!" It took him about three tries, but each time the crowd got a little quieter. Finally, when he had everyone's attention, he began.

"As you may or may not know, our 'beloved' Secretary of Defense, is making one of his famous fact-finding visits to Southeast Asia tomorrow. As you may or may not know, he is apprehensive that by some miracle a MiG might exceed its combat radius (by a factor of about five) and pull a sneak attack on him. For that reason, a fighter escort has to be provided for him everywhere he goes in South Vietnam. My flight has been given the dubious honor of escorting the Defense Secretary when he honors us with his presence tomorrow. All I want to do is announce to you that I'm going to shoot the son-of-a-bitch down. Thank you."

With that, my friend turned around and sat back down. The silence was deafening. A colonel who was sitting at the end of the bar made for the quiet pilot. He called him by his first name, and put his arm around the pilot's shoulder.

"Son, you can't say things like that," he said. "You'll get into big trouble. It's just the alcohol talking."

"No, sir. It's not the alcohol, it's me, and I'm going to do all of us a big favor and shoot the S.O.B. down," the pilot said.

"No, you're not," the colonel said. "You're grounded." He then called the bartender over and had him call the Air Police. When they came, he instructed them to keep someone with the F-4 pilot until further notice and then made the call to ground him.

The Defense Secretary made his morale busting—sorry, morale *boosting*—visit the next day as announced. He evidently learned everything he needed to know about the war and was gone in about thirty minutes. I suppose it doesn't take a politician as long as it does the rest of us to learn what he needs to know.

The next day, the Secretary of Defense left Southeast Asia, and the fighter pilot was ungrounded. Although he was never informed of the incident, the next time the Defense Secretary visited, he wouldn't have had to worry.

The Phantom pilot and his backseater were dead.

Why me?

Cam Ranh Bay Air Base, RVN, 1966

Over nine months had elapsed since our Fighter Wing had arrived overhead our new "home" and looked down at the long, sandy peninsula. There wasn't much to look at. A few Butler Buildings were connected by dirt roads scraped through the sand. To the west of the buildings lay the big flat aluminum-mat ramp. Tall aluminum revetments to help protect the F-4s from attack were dividing the large ramp into two-plane sections.

Connected to the ramp by narrow taxiways was the long and narrow north-south runway. It looked even longer because of its very narrow width. If I remember correctly, it was only about one hundred feet wide. Recall that it was nothing more than panels of aluminum matting hooked together and laid out on the sand, which had been somewhat stabilized by a thin layer of asphalt.

To the west and southwest were hills abruptly rising up to several hundred feet. To the south was the Bay of Cam Ranh. There were lots of supply vessels docked and sitting anchored in the blue bay; that was where I found out the U.S. Army had big boats. To the north was more water, then several miles up the coast, a U.S. Army supply base and airfield called Nha Trang. East of the base building area was one of the most beautiful beaches I have seen. The beach was still the same.

Everything else had changed. There were newer buildings under construction. Across the aluminum runway to the west, a brand-new concrete runway was almost finished. The nearly new F-4s that we had flown there from Florida with such great anticipation were now ragged and worn looking. The once-new camouflage paint had peeled back from areas with more exposure to the swift airflow. There were thick aluminum patches on most of the jets. There were not only patches for battle damage, but they were also attached to areas that had been shown to be more subject to stress fractures, such as just inboard of the wing-fold areas.

I was thankful that I had brought new Phantoms in. I was glad that I was not just arriving to fly the war-weary birds that we would leave behind. I was amazed and grateful that these jets had held up so well under the punishment that daily combat missions carrying heavy loads imposed upon them. They were fantastically durable, a great tribute to their manufacturer, McDonnell Aircraft at St. Louis, Missouri.

The "patches" on the pilots weren't as visible as those on our jets. You had to look closely to see any signs of the "stress fractures," but there were hints. Eyes that had once sparkled when laughing at a joke or prank had dimmed slightly. The skin around the eyes now looked dark and drawn, not taut and healthy. Sometimes there was just a little slump to the shoulders on the walk toward the jets. We were getting tired. Most of us had lost weight but our slimness was not healthy looking. It more resembled the weight loss from some prolonged illness.

The dead and missing were *not* mentioned. They just weren't *there* anymore. Fresh new faces replaced them. Some didn't last long. One young new pilot only lasted three short days before he joined the others, not *there*.

I was kidded sometimes because at one point I held the Wing "record" for getting my F-4 hit on the most missions. It was nine times. I had gotten my right fuel drop tank blown up twice. They accused me of "leading with my right." But I didn't mind the kidding.

I have dead friends who were hit only once.

A picture on my wall right now causes me to pause on occasion with wonder. There are four people in it. I and another pilot from MacDill had flown an F-4 to Laughlin Air Force Base at Del Rio, Texas, for a static display in 1964. The Wing Commander and a pilot from Laughlin are standing there with us.

The pilot with me from MacDill later suffered a direct hit from a surface-to-air missile over North Vietnam. His backseater was killed outright, but he ejected and spent several years as a prisoner of war at the "Hanoi Hilton."

The pilot at the left of the photo was a friend of mine and the future POW's. He became "Sidewinder 21." You might remember that previous story. I last saw him alive at Tan Son Nhut Air Base when I had first arrived in Vietnam.

Of the three young pilots in the photo, one dead, one POW, and me. Why me?

Lucky number 205

Cam Ranh Bay Air Base, RVN, 1 October 1966

I knew time was running out. One by one, our squadron pilots were getting their assignments. One by one, they would return from a flight and be informed that they had just flown their last combat mission. It was a wing policy to not tell you that you were flying your last flight in Vietnam. I think it was something of a superstition. I knew that it wouldn't be long. I had even tried to extend my tour in Vietnam, so I could continue hunting MiGs, but my request was denied. I flew every air-to-air mission that I could to the north, but it just never happened.

I taxied in, and there they were. The squadron commander's jeep was waiting, and they already had the water buckets out and were shaking up the cheap champagne. I really had mixed emotions. I wanted to go home, but I felt somewhat unfulfilled. I had flown two hundred and five missions, four hundred and twenty hours of combat time. Most of them had been air-to-ground missions. We hadn't received enough air refueling support to begin flying into North Vietnam until about six months after arriving. Probably a third of my missions had been in the south. Another third had been in *that* country, where, according to our honorable politicians, we weren't flying. The remainder were in the North, or just off its shore, providing MiG cover for some of our more vulnerable friends. But suddenly, it was over.

I knew that I had changed since arriving in Southeast Asia, although some of my friends had not seemed to change at all. I felt anger and animosity toward the enemy and what they represented, but I did *not* enjoy killing them. I don't know a single pilot who did. I will confess however, that I absolutely loved the flying! I reveled in the high speeds and extremely low altitudes. I liked pulling the Gs.

I thoroughly enjoyed pushing the Phantom to her limits and feeling her respond. She was a truly great airplane. She was big and not particularly attractive to most beholders, but to me she was beautiful. Her strength and power saved me on more than one occasion. She was extremely effective and feared by the enemy. It was a love affair that still continues.

I was hit by enemy ground fire on nine different missions, but never knocked down. Twice I returned from missions with one engine out. Even after I had been hit several times, I still didn't worry about getting killed. My confidence never waned and actually grew as I became more experienced. I felt at home in the cockpit. Unlike some, I never had to look for false assurance in a bottle, or try to escape into one. I felt pity for those who did.

I think I was born to be a fighter pilot. At least I feel that I was blessed with the right kind of eyesight, the right reflexes, the right feel for the fighter. It wasn't work for me to fly them. It was sheer pleasure and intense enjoyment.

I read once that being a fighter pilot is the most demanding and dangerous occupation in the world. It requires the operation of a dynamic machine with tens of thousands of horsepower at tremendous speeds and high forces of gravity. It's a three-dimensional world where upside down is sometimes best. It's a realm where going as fast as possible, as low as possible, is sometimes safest. It's being in control of a demanding beast, requiring the simultaneous movement and coordination of hands, feet and eyes, all while turning, rolling, diving, and climbing with your body sometimes weighing eight times its norm.

Your two-hundred-pound, six-foot frame is trying to be four-feet tall, and it weighs sixteen hundred pounds. Your blood, eight times as heavy, tries to rush from where it's needed for vision and ultimately consciousness. You have to strain to keep the blood up where it belongs. Sometimes you begin to lose it, and your vision starts to "tunnel." You have to strain even harder and ease off the back pressure on the control stick to stay conscious. You're in control, but you know that your plane will kill you in an instant if you make a mistake.

But you wouldn't trade it. That adrenalin high is like nothing else in the world. The challenge of flying and mastering such a machine is rewarding, exhilarating, and unmatched! *Now* you add lots of guys shooting at you!

Who wouldn't love it?

You would if you were, *Once A Fighter Pilot.*

About the author

Brig. Gen. Jerry W. Cook, USAFR, Ret., has been flying for more than thirty-eight years. After ten years in the U.S. Air Force, he left in 1967 to fly for Pan American World Airways. Cook flew for sixteen years with the Arkansas Air National Guard, retiring as a brigadier general in 1986. He lives in Little Rock, Arkansas, and is a corporate pilot for Stephens Group, Inc.